ENGINEERING 50 YEARS
FROM TODAY

THE FOUNDERS SERIES

The Founders Series publishes books on and about Purdue University, whether the physical campus, the University's impact on the region and world, or the many visionaries who attended or worked at the University.

ENGINEERING 50 YEARS FROM TODAY

A Purdue Engineering Community Perspective

EDITED BY
ARVIND RAMAN

Purdue University Press
West Lafayette, Indiana

978-1-62671-283-6 (paperback)
978-1-62671-284-3 (epdf)
978-1-62671-285-0 (epub)

Cover images: Approach of the SpaceX Dragon Commercial cargo craft: ISS033-E-011328 (10 Oct. 2012)/ https://images.nasa.gov/details/iss033e011328; Venus view: Vanit Janthra/iStock via Getty Images Plus; Space Background Star Nebula Cosmos Texture Sky Cosmic Astronomy Black Universe Galaxy Outer Deep Dark Starry Light Night Abstract Dust Planet Sparkle Winter Backdrop Earth World Milky Way Astronomy: wing-wing/iStock via Getty Images Plus; Panoramic view of the Earth. Space: buradaki/iStock via Getty Images Plus; Panorama sky with cloud on a sunny day: Tanarch/iStock via Getty Images Plus; 3D illustration of a DNA molecule with sparkling effects symbolizing complexity and genetic diversity. Futuristic concept of genomics, precision medicine, and computational life sciences. H: TanyaJoy/iStock via Getty Images Plus; ITER CA SuperWide Shot / Tokamak, International Fusion reactor 4K: John D/iStock via Getty Images Plus; Server racks full of routers, switches, and servers aligning on both sides of an aisle of a data center. Illustration of the concept of cloud computing and infrastructure as a service (IaaS): Dragon Claws/iStock via Getty Images Plus; AI, Circuit board: Andy/iStock via Getty Images Plus; Saturn with stars in the background: MarcelC/iStock via Getty Images Plus; Neil Armstrong Statue, 2023_RR13660.jpg: Purdue Brand Studio.

To all Boilermaker engineers past, present, and future

CONTENTS

*Essays from Mrs. Lehman's class

PREFACE

As this book about the future of engineering is being written we are entering the final few months of a yearlong celebration of the 150th anniversary of the founding of Purdue University's engineering program. In September 1874 President Abraham Shortridge published the first engineering-wide curriculum, approved by the faculty, that consisted of three years of a common curriculum and a fourth year of specialization in one of four majors offered at that time. Our historical archives indicate that in the fall of 1874 William Morgan was the first engineering student enrolled at Purdue out of 39 students at that moment in time and was taking a class taught by William Morgan, the chair of Engineering and Mathematics. September 1874 is thus seen as the point of origin of Purdue's entire engineering program, even while the engineering departments and the College of Engineering as a whole were established much later.

Since then, the Boilermaker Engineering brand has shone brilliantly thanks to generations of Boilermaker engineers who have walked in through the wide doors of opportunity of Purdue engineering always striving for that next giant leap. Today we truly stand on the shoulders of Boilermaker engineer giants. Some of the biggest giant leaps globally in the past 150 years have had Boilermaker engineers leading the charge behind them: the Golden Gate Bridge, the Hoover Dam, human space exploration, nitrogen fertilizer, MOSFET, Kevlar, insulin, pacemakers, television, wireless communication, pattern recognition, and so much more.

In the most important sense, the future of engineering has much in common with the past in that the enduring purpose of engineering is to innovate creatively and find solutions that benefit people and society. The rapid advances in science and technology are giving engineers ever more raw material with which to creatively do their work, even before the science is fully elaborated. This people- and society-focused creativity gives engineering its distinctiveness in solving problems and realizing opportunities that improve health, human knowledge, quality of life, energy, space exploration, global security and economic productivity and sustainability in agriculture, manufacturing, transportation, and logistics.

The famous Yogi Berra quote "It's tough to make predictions, especially about the future," certainly applies to this book. However, our essays approach the difficult problem of predicting the future of engineering the same way all difficult engineering problems are addressed–starting with the fundamentals.

For example, how energy is provided to society is a fundamental issue. In this regard the engineers of the next 50 years face one of the most profound and difficult engineering challenges in history: how to increase manifold humanity's access to energy while making it carbon-free, secure, and affordable. Whether by harnessing the power of stars (nuclear fusion), small modular reactors, advanced nuclear reactors, or vastly increasing carbon-free grid-coupled energy production and storage, engineers in the future will strive to find solutions to this daunting challenge. The rapidly converging critical mass of technologies makes the practical engineering of wonderous future energy systems a good bet, much as the engineering of critical mass technologies achieved heavier than air flight in the early 1900s. The essays in this book explore these and many more fundamental issues that will be critical to the future of humanity.

One final thought we are really excited about is how the teaching and learning of engineering will change in the next 50 years. How will engineering education embrace artificial intelligence and make it easier for engineers to find creative innovations to solve problems for people and society? What will the role of future engineering educators look like? What will instructional labs of the future look like? As educators, we must develop increasingly specialized technical subject matter along with a universal sense of responsibility, wonder, and awareness so that engineering benefits as many people as possible. Given the importance of engineering to the future, this book addresses the topics mentioned above and much more. Please enjoy the essays from the many thoughtful authors who provide rich perspectives on the future of engineering and why today is one of the most exciting times to be and become an engineer, especially a Boilermaker engineer!

1

SUPERENGINEERS

A Student's Thought Process

KEVIN BOES

Kevin Boes is a PhD student in mechanical engineering at Purdue and researches supersonic flows at Zucrow Labs. He is serving a two-year term as the student on Purdue's Board of Trustees.

As I sat in my windowless office awaiting a computational simulation to finish amid a fortress of screens, I glanced at the quadruple-monitor setup towering in front of me. Each screen overflowed with information—email threads on one, simulation outputs on another, a to-do list scattered across two apps, and a half-read research paper pushed to the corner. A notification pinged on my phone: "Hey, do you have a minute?" My watch vibrated, reminding me to stand. My adviser pinged me, asking about how the journal article was coming. Overstimulated, I defaulted to the next form of stimulation: I grabbed another cup of coffee and scrolled social media while listening to my favorite music, forgetting about the PhD work.

This is the modern reality of solving problems as engineers and scientists: a whirlwind of tools promising productivity but often delivering distraction. I spend half my day toggling between tasks, constantly on the verge of deep work but pulled away by a cascade of pings. Each tool promises efficiency, but collectively they create an overload—of information, stimulation, and decisions.

I returned to my desk and observed the squiggly lines on my simulation residual plot, hoping for convergence. The lines spiked rapidly, indicating a failed simulation. Nooooo! Why? Was it the boundary conditions? How am I modeling turbulence? What even is turbulence? Why am I running this simulation? How will this simulation help someone? Wait, why am I even doing a . . . ?

Pause. Amid the questioning, I stepped outside to decompress. The vast Montana skies and rugged mountains of my roots taught me the value of getting lost—both in nature and in thought. But Indiana feels different. A wind shear force scraped my face immediately as I stepped outside. I looked into the distance: corn.

I took a few steps, each one pressing against the cold air. It was too cold for this, but I gritted my teeth and pushed forward. The simplicity of walking stood in stark contrast to the complexity I had left behind in the lab at Purdue. As I moved, the blood flow started, and my thoughts began to organize themselves.

For a moment, I thought back to the simplicity of childhood in Montana, a time when the biggest challenge was getting up for morning football practice. Now, with a broader perspective, the challenges felt immeasurably larger: harnessing the evolution of artificial intelligence (AI), curbing carbon emissions to save the planet, or even making humanity a multiplanetary species. As someone driven to contribute, I found myself wondering what it would take to tackle these challenges. New tools? New ways of thinking?

The tools we use as engineers are powerful beyond belief. Today, I can obtain gigabytes of simulation data of supersonic flow in minutes, consuming the equivalent electricity of 150 loads of laundry. AI tools such as ChatGPT can solve homework problems instantly, something I could only dream of as an undergraduate. At the same time, the internet floods us with endless information that often blurs the line between fact and fiction, such as debates over what is healthy to eat and TikToks claiming the discovery of the secret to immortality. This acceleration of technology begs a deeper question: In a future saturated with information and distractions, how can I, as an engineer, rise above the noise to create meaningful impact?

My answer became clearer with every step. I believe it is not just about mastering the tools but also developing agency. Future engineers must be more than technical experts. It is almost like future engineers need to be superengineers—agents of vision and purpose capable of thriving amid the chaos of overstimulation through clarity and intentionality. This clarity, however, is becoming increasingly difficult to achieve.

These days, I catch myself doomscrolling absentmindedly through articles and social media, filling work gaps with fleeting dopamine hits. The future, I imagine, will be even more extreme. Augmented reality glasses will overlay data on everything we see. Notifications will be tailored to our every thought, feeding us answers before we even know what questions to ask. The future suggests that algorithms will constantly churn in the background, influencing every aspect of our lives.

It seems that engineers will face a new kind of problem—one not of scarcity but instead of abundance. The ability to sift through noise, distill clarity from overload, and focus on what truly matters will define those who make a lasting impact. I saw this challenge embodied during my time at SpaceX in the summer of 2022 in Texas.

On my first day after I parked with a stunning view of the upright Starship rocket and the partially built launch mount gleaming in the distance, my boss greeted me while simultaneously listening to a debrief of the recent successful Starship rocket landing. The next challenge was launching the full orbital system—Starship and its booster—by July 2022, as the countdown clocks around the facility reminded everyone.

I worked out of a house converted into an office, where ideas often sounded ambitious if not outright impossible. I remember the talk about using guide rails to catch the booster: "Like chopsticks," they said. Yeah, sure. Fast-forward two years to 2024, and they actually did it. Amazing. I suppose sometimes it *is* rocket science.

The engineers at SpaceX are truly elite, some of the highest-caliber problem solvers I've met. And many of them are Purdue alumni! They operate with boldness and clarity, guided by a shared vision of making life multiplanetary. Their approach is a lesson in "first principles" thinking: break down a problem to its raw essentials, then build your solutions from the ground up.

I recall explaining to my boss all the fascinating details I was learning about hydraulic actuators from my research while designing a weldment. He stared at me blankly and asked, "How does your research contribute to making life multiplanetary?" I paused, caught off guard. "Just order the actuator. It's a six-week lead time," he said, "and tell them we need it now." At SpaceX, if a task doesn't directly advance the mission of making life multiplanetary, it gets deleted. This clarity of focus is a blueprint for how to thrive in abundance: block out the noise, prioritize ruthlessly, and iterate rapidly.

SpaceX is arguably one of the most innovative companies in history, and its engineers seem to operate on a different wavelength. They represent the engineers of the future: superengineers. To tackle the next generation of challenges, we need more superengineers.

Feeling inspired, I pulled out a pen and paper from my coat pocket as I passed the bell tower on campus. There was something grounding about the act of sketching an idea—no screens, no algorithms, just the simplicity of a blank sheet.

I began to imagine what a superengineer might look like. "Begin with the end in mind," as Stephen Covey says. Out of curiosity, I pulled out my phone and asked ChatGPT to generate images of an engineer. It gave me predictable results: people glued to screens with diagrams or wearing hard hats and holding clipboards on construction sites. But this wasn't the vision I had in mind.

I refined my query, asking ChatGPT to imagine a superengineer. After I refined the prompt 10 more times and asked it to make the engineer more "super," the result was much better. It depicted a sleek, robotic humanoid with the power of the galaxy in its hands—literally. The figure balanced two galaxies while flying hypersonically alongside rockets, a neon-lit megacity glowing below. It was like a scene from my favorite childhood hero, Iron Man: bold, innovative, and unafraid to tackle impossible challenges. This, I thought, was closer to what I aspired to be: an engineer who balances technical mastery with creativity, vision, and the courage to push boundaries.

As I reflected on this ideal, I found myself walking through the halls of the Mechanical Engineering building, where relics of the past line the walls—pulley systems, drafting tools, and the large mechanical clock that somehow still ticks. These tools, symbols of a more manual age, amazed me with their ingenuity but also highlighted how much has changed. My grandpa often shared stories of his days as a geological engineer in Colorado, meticulously hand-drawing topological maps and crafting physical subsurface terrain models—work that required precision, patience, and a steady hand. Today's engineers, by contrast, are knowledge workers, relying more on their minds than their hands. The shift is clear: less hands-on, more brains-on.

I envision a future where mastery over our neurological functions becomes invaluable. The quality of work will no longer be measured by the precision of physical drawings but instead by the brilliance of the ideas behind them. Advanced human-machine interfaces, perhaps nearly telekinetic, will enable seamless collaboration with digital twin models. As tools such as CAD and drafting become increasingly automated, the enduring skills such as reading, writing, and thinking—the deep work—will remain timeless.

As I turned back toward the lab, the cloud cover vanished, and the sun started to shine. It is almost like the weather had changed from freezing to manageable in seconds. Amid the brightness, I imagined what Purdue's role might be in this bright future. For 150 years, Purdue engineers have embodied resilience, practicality, and work ethic. But the next 50 years demand more—not just in scientific breakthroughs but also in how we bring those breakthroughs to the world.

Passing by the Neil Armstrong statue, I couldn't help but feel inspired by the legacy of those who came before me. It amazes me how universities such as Purdue have been instrumental in advancing technology, especially post–World War II. The increased collaboration between academia, industry, and government gave rise to such achievements as the semiconductor revolution, sustainable agriculture to address global food security, and even walking on the moon. Purdue, with its legacy of producing pioneers such as Neil Armstrong, exemplifies this spirit of partnership and innovation.

As a 2022 graduate from Purdue, I am incredibly proud to be part of a tradition of innovation and impact. I envision a Purdue that not only leads in knowledge dissemination but also takes on the duty of commercializing discoveries to better serve society. This means creating a system where ideas don't just stay in academic journals but also become products and solutions that address real-world problems: clean energy technologies, autonomous systems, and advanced materials that improve daily life.

I believe in the opportunity for Purdue to strengthen its role as a hub for entrepreneurship that rivals the coasts, turning West Lafayette and Indiana into a hardtech corridor. Achieving this vision will require the collective effort of Indiana's people, leveraging our shared strengths of work ethic, resilience, and attitude. By fostering a seamless ecosystem of research and commercialization, Purdue can continue to lead not just in developing new technologies but also in ensuring that they reach the people who need them most, driving progress for generations to come.

When I reached my office after this insightful walk, the familiar glow of the monitors and the steady hum of the machines greeted me. But this time, I didn't feel overwhelmed. A light bulb switched on in my mind—I realized that I didn't need the simulation to answer my research question. The trends from simple equations were enough.

It's odd how sometimes I have the best ideas while *not* working—walking, showering, driving. This realization tied into a broader truth I had been grappling with: the future of engineering is not just about mastering the tools, knowing the perfect equation to use, or even the sheer number of hours worked; it's also about cultivating a mindset that values creativity, clarity, and purpose. I started my walk questioning why I'm pursuing a PhD, but by the end I understood: I'm here to unlock the full potential of the human mind through not just what I discover but also what I uncover about myself. And I'm ever grateful that Purdue is there to challenge me to do exactly that.

2

THE IMAGE PROCESSING MAGICIANS AT PURDUE AND THE FUTURE OF VISUAL ENGINEERING

AL BOVIK

Dr. Al Bovik previously served as the director of the Laboratory for Image and Video Engineering, at The University of Texas at Austin. He is now Provost's Chair Professor in engineering at the University of Colorado Boulder.

There was a beautiful but sweaty day in the backcountry of Belize in March 1996, a few hours from Belize City, where I was to attend the annual Institute of Electrical and Electronics Engineers (IEEE) Signal Processing Society Workshop on image processing but had arrived early to go exploring. A colleague and I left the workshop to hike the land of jungles, jaguars, and the Jabiru stork. Along the way, we stopped at a roadside stand to buy some water. The proprietor was an operator named Cesar who wondered why we were there. He said that a fellow had preceded us. "This guy—straw hat, beard, glasses—do you know him? Is he police? A fed? Something about him. He asked a lot of questions." We shook our heads. Clearly, Cesar had something to hide.

After driving down a dirt road deep into the jungle, heaving our bags we headed toward the path winding down to a river and a circle of huts to meet the hut keeper and his ghastly dog with half a face (courtesy of a local jaguar). It was exciting and mysterious, and we'd have a lot to talk about at the workshop! However, just as we set foot on the jungle path, out of the forest parades the distinguished professor Edward John Delp of Purdue University, world-famous image processing engineer, bearded face topped by straw hat, smiling through his spectacles, his wife Marian by his side. Ed Delp had already been to the jungle ahead of the rest of us.

More than a decade before that, I was an engineering student at the University of Illinois at Urbana-Champaign (UIUC) in the 1970s and early 1980s, getting my BSEE in 1980, and then my graduate degrees under the venerable Thomas S. Huang and David Munson. Dave is currently the president of the Rochester Institute of Technology in New York. Tom passed away a few years ago, may his great soul rest in peace. Tom, whom many regard as among the greatest image processing engineers ever, was a professor at Purdue from 1973 to 1980, where he played an essential role in building the image processing group there. I learned image processing from Tom and Dave, and it remains my passion and vocation today, even as it has become a ubiquitous, transformative technology.

During those eight years at UIUC, I was vaguely aware of a similar university across the Illinois-Indiana border in West Lafayette, one of the top engineering schools in the world, a land-grant state university like UIUC. It was less than 100 miles away, yet in my eight years at university I never found my way to Purdue. After all, I only owned a bicycle, and while I occasionally took 50-mile rides through the endless cornfields (one target was a tiny town called Flatville), the 200-mile round trip was too much.

I joined The University of Texas at Austin in 1984, teaching digital image processing and computer vision. These topics weren't used much in the 1980s other than in high-dollar military and medical imaging applications, but it was apparent that one day they would be really important. Conferences and workshops on those topics were well attended and robust, full of opinionated people. It was at events such as these where I would meet fellow young professors of great talent and passion, such as Rama Chellappa, Aggelos Katsaggelos, and, of course, Ed Delp. I remember sitting on a bench sipping beers with Ed and five other young image processing stars and talking about high-flying topics such as nonlinear digital filters, order statistics, and mathematical morphology.

I learned that Purdue had one of the strongest image processing groups in the world, led by young professors such as Charlie Bouman, Jan Allebach, Leah Jamieson, Ed Coyle, and Ed Delp. It was (and is) a remarkable group of amazing researchers who worked well together.

Charlie's work greatly advanced the field of medical tomography. He helped establish the field of computational imaging and was the driving force behind the creation of the journal *IEEE Transactions on Computation Imaging*, serving as its first editor in chief. He has always been a great public servant and previous to that was editor in chief of the journal *IEEE Transactions on Image Processing*. I am sure he feels that his greatest accomplishment, however, was bringing up his amazing daughter Katie, who was part of the team that captured the first picture of a black hole!

Jan found tremendous fame in the printing industry; his digital signal processing inventions are used in laser printers around the world, helping them produce crisp, high-quality prints with great efficiency. His most famous inventions used principles of visual perception to create halftoning algorithms, which print versions of real-world pictures that are made of just black-and-white dots in such a way that makes them look like they have all shades of gray and appear realistic. The printing industry loves this, and your laser printer probably used Jan's algorithms.

Ed Coyle helped found the field of statistical image processing and invented the concept of stack filters, which were provably optimal for image denoising. He and Leah Jamieson went on to become great contributors to engineering education and won the Gordon Prize from the National Academy of Engineering for their innovations. Leah, meanwhile, served as not only the dean of the School of Engineering at Purdue but also president of the IEEE, the world's largest technical organization.

Ed Delp has been a cosmic force in the signal processing community for better than four decades. At any IEEE Signal Processing Society conferences, wherever it was in the world, globetrotting Ed could be found at the center of the hall, with a crowd of students gathered around or flying from poster to poster like a live electron. He has always been one of the Signal Processing Society's greatest orators because of his humor and precise targeting of the important problems and his ability to make complex topics simple. Early in his career Ed invented a concept called block truncation coding, a very effective image coding algorithm that is also trivially easy to implement. For these reasons, block truncation coding was used by the National Aeronautics and Space Administration to communicate back to Earth all the images of the surface of Mars taken by the cameras on the Mars Pathfinder robot, affecting hundreds of millions of people with the wonders of the Red Planet. Ed also helped invent the field of image forensics. Some of his work has involved digitally analyzing pictures to determine which exact camera from among millions took the pictures. For these amazing contributions, Ed has received every major award for his research from the IEEE Signal Processing Society.

Throughout my professional life, the group of image processing researchers at Purdue has always been at the center of innovation, service, and education, representing a huge impact on society. Today, people view images in digital displays everywhere, including the phones in their pockets and their increasingly huge televisions, tablets, and monitors. Nearly every branch of science has benefited by being able to visualize their results in, for example, medical imaging and astronomy. But the visual engineering transformation of society has just begun!

The tomorrow of image processing, over the next 50 years, promises to be just as transformative if not more so. Processing images with increasing powerful computers will continue to play a centric and critical role. I believe that some of these technologies will serve to bring us closer together and will provide us with greater safety. The day is not far off when friends will be able to communicate with each other visually, as if they were in each other's presence, in full realistic 3D in whatever earthly (or even unworldly) environment they choose. They will be able to do this wherever they might be, unshackled by cables and monitors and only requiring a light pair of glasses. These interactions will allow users to walk anywhere they like, be it the wilds of

Alaska or crossing the street at Times Square. I think we will reach this in less than 10 years, judging from progress at several tech companies I work with. Another revolution that is coming is interconnected automobiles. There are already driverless cars zipping about cities using many cameras, massive amounts of image processing, and geolocation to navigate, while humans sit in the back seat reading their emails. Soon, these vehicles will be wirelessly interconnected and communicating with each other, not only avoiding collisions but also planning collectively optimized traffic patterns. The day will come when autos whiz about at much higher speeds, perhaps without any traffic lights, with little slowing of any vehicle between pickup and destination. This may take another 20 years, but it is coming. Of course, there is that little worry about pedestrians, which suggests that the time is coming for better ways to cross the street. Naturally, these vehicles will be driven and organized by massive artificial intelligence (AI), which after all represents the ultimate optimization engines.

Increases in wireless bandwidth are made possible by using higher-frequency carrier signals. This technology has been arriving for some time now in the form of millimeter wave/gigahertz wireless signaling but is limited by the distances these signals are able to travel and their inability to pass through solid objects. Given the demands of increasingly sophisticated video users, this may be solved by ultra high-frequency signaling and extreme densification of antennas that take up very little space. This process has just started. Wireless relay stations supporting extremely dense traffic are evolving into smaller, cheaper, and lower power and will be available everywhere inside and outside of homes, office buildings, grocery stores, and more. Video, the biggest data, already occupies about 80% of all internet capacity. Users will demand huge amounts of visual data, often immersive and in 3D, regardless of where they are. The advanced augmented reality scenario I mentioned above is a good example, requiring much larger and faster videos so that near-perfect realism is achieved and users can be "on the field" when Shohei Ohtani makes his next big hit!

It should be no surprise that I think that robotics will be an immense role in daily life within the next 50 years. Virtual AI techniques are already creating highly realistic visual avatars that give voice to language generated by large language models such as ChatGPT. Likewise, physical robotic technologies are advancing at a tremendous pace. There are robots that can do most industrial tasks (such as Amazon's almost fully automated distribution centers), and there are humanoid robots that can walk, jump, and dance and have increasingly realistic skin! Within a couple of decades, humanoids will be among us on the streets and perhaps in the office and the home, combining these developments in naturalistic and seemingly sentient forms. This will present tremendous societal challenges, of course, deeply affecting people's lives in many ways. As these synthetic beings become indistinguishable from ourselves, many people will be affected by the uncanny valley wherein robotic behavior falls a bit short of perfectly real, making robots detectable, queer, and bizarre enough to engender negative reactions. Asimov, anyone? Isaac Asimov supplied many plausible scenarios that we may encounter even if his positronic brains miss the mark of reality.

In the early part of the new millennium, Ed Delp visited The University of Texas at Austin, where he gave a prescient lecture and met with, entertained, and educated all of my students. I returned the favor in 2023, giving a talk on video quality that was well attended by many enthusiastic bright minds. I visited Ed's laboratory, met his students and grilled them on their work, and sat and chatted with Ed in his office for quite some time. Ed's office is filled with four decades of books, notes, and engineering artifacts, including a stack of what looked like every cell phone ever made. We talked about the state of the field and the steamroller of deep learning that was changing everything and even discussed retirement, a strange topic for two engineering educators who haven't given the topic much thought. But time marches on. We recognized that we both still have a lot to contribute. Ed is still revolutionizing the field of visual nutrition (using AI to help analyze what's on our plate and make recommendations), and we both love being in the classroom. But fewer of our colleagues are still laboring or are even with us, and the field has changed so much because of deep learning that it is now barely recognizable. However, neither of us are the type to fade into irrelevance like James Hilton's Mr. Chips or Willa Cather's Professor St. Peter, and I think we will both ride the new wave of AI into its future in image processing (who better than us, after all?) and eventually into the sunset.

3

150 YEARS OF PURDUE ENGINEERING—AND THE NEXT 50 YEARS?

How Fortunate That I Have Lived the Last 70 Years!

WENG C. CHEW

Early in his career, **Weng C. Chew** programmed with the early punch card IBM mainframe computer at MIT and then became avidly interested in electromagnetics due to its impactful nature and its wide range of validity. He then spent 32 years at the University of Illinois and eventually became a distinguished professor at Purdue University.

I am glad to have lived in the last 70 years, the period of the most rapid whirlwind technological change in human history. I grew up in the developing world in Malaysia with no running water, and I first saw a radio in our neighbor's house when I was six years old. As a teenager, I learned how to build my first vacuum tube and transistor radios. I lived through the era of Sputnik and later the moon landing by Neil Armstrong.

Long before my time, however, several monumental events happened in the 1860s. The land-grant university act was established in 1862, the assassination of Abraham Lincoln happened in 1865, and also in 1865 James Clerk Maxwell's equations were published by the British Royal Society.

The birth of Purdue engineering and the land-grant universities in America happened around the same time.

Therefore, the birth of Purdue University and many land-grant universities roughly coincides with the beginning of 150 years of electromagnetics technologies. These technologies were primarily governed by Maxwell's equations. This set of equations, consisting of four major equations, set the stage for engineering science for the next 150 years even though Maxwell had a relatively short life span (1831–1879).

THE FIRST 50 YEARS

Maxwell's equations, in the notation used by Oliver Heaviside,[1] are

$$\nabla \times \boldsymbol{H} = \partial \boldsymbol{D}/\partial t + \mathbf{J}$$
$$\nabla \times \boldsymbol{E} = -\partial \boldsymbol{B}/\partial t$$
$$\nabla \cdot \mathbf{B} = 0$$
$$\nabla \cdot \mathbf{D} = \varrho$$

Simple though the equations look to the experts, they have been the harbinger of modern electrical and electronic engineering.[2]

In the beginning, Maxwell's equations were the foundational theory that gave rise to electrical circuits theory such as Kirchhoff's circuit law and Kirchhoff's voltage law. Later, the equations yielded Faraday's law (1831), Ampere-Maxwell's law (1855), and Gauss's law for magnetic flux as well as electric flux. In the early days (1820), these laws were motivated by experimental measurements. It was found that when current flows in a loop, it gives rise to magnetic flux, or

$$\nabla \times H = J$$

This law is known as Ampere's law.

Then it was discovered that time-varying magnetic flux gives rise to the electric field that gives rise to voltage. This law was known as Faraday's law. It was realized that this voltage can be used to drive electric current through a wire even though the wire has small resistance. The fact that a current flowing through a small resistance gave rise to a voltage drop, known as Ohm's law. When voltages are summed over a closed loop, they add up to zero, giving rise to Kirchhoff's voltage law. Then it could be shown that there was divergence of $J = 0$, assuming that the frequency was very low. This law was known as Kirchhoff's current law. These two laws—Kirchhoff's voltage law and Kirchhoff's current law—are the fundamentals of circuit theory, which has been the driver of electrical engineering for over 100 years now.[3]

One of the early technologies that emerged from electromagnetics was telegraphy. Telegraphy was used to transmit information long distances via the use of telegrams and Morse codes. Telegraphy also allowed the transmission of information using telegraph cables. As early as the 1800s, these cables were laid under the ocean from Britain all the way to Hong Kong on the rim of the South China Sea. With the turn of a switch, Queen Victoria could send a telegraph signal all the way from London to Hong Kong in 1871.[4]

What is more interesting is that the original four equations of electricity and magnetism were incomplete. For instance, they could not explain why electric current can flow through a capacitor. To complete the equations and make them consistent with circuit theory and charge conservation, James Clerk Maxwell added a displacement current term to Ampere's law:

$$\nabla \times H = \partial D / \partial t + J$$

Only with this modified Ampere's law, also called the generalized Ampere's law by some authors, was electromagnetic theory complete. One can show the emergence of wave theory from the completed electromagnetic theory. These equations have been shown to be valid from atomic length scale to galactic length scale. They can be used to calculate the interaction of an electromagnetic field with protons and electrons inside an atom. Also, many of these are subatomic particles whose sizes are much smaller than the atoms; hence, it is accepted that electromagnetic theory is valid down to subatomic length scale.

In addition, many of these subatomic particles have spins, with dipole moments like that of a small magnet. These dipole moments have been calculated using theoretical physics methods. These dipole moments are validated to high precision, showing the validity of electromagnetic theory.[5]

Electromagnetic theory has also inspired special relativity, implying that the speed of light in a vacuum is a universal constant. This also implies that Maxwell's equations remain the same irrespective of what inertial reference frame the measurement was done to validate the theory.

Therefore, electromagnetics theory has been validated from subatomic length scale to galactic length scale. At present, the uses of electromagnetic theory have ranged from nanolithography to intergalactic communications.[6]

THE BIRTH OF ELECTRICAL ENGINEERING AND COMPUTER TECHNOLOGY

Maxwell's equations allow engineers and scientists to manipulate the flow of electrons, which is electricity. This was the beginning of electrical engineering. Electricity was then generated by voltaic cells that used chemistry to generate electricity. Later, electric generators were invented to convert mechanical motion into electricity. This allowed engineers to harness the power of steam engines, turbines, and windmills to generate electricity. Electricity operates in reverse and can power a motor.

Electricity was then used to power telegraphy, which requires switching technology to generate Morse codes. Switching technology requires nonlinear circuits, and the vacuum tube was one such device. Electronics came about after vacuum tubes. Vacuum tubes, with their ability to rectify and amplify signals, gave rise to the birth of radio engineering. Vacuum tubes consumed too much power, which later motivated the invention of energy-efficient diodes, transistors, and nonlinear circuits as rectifiers and amplifiers. Transistor electronics were too noisy in the beginning, but with advances in materials, high-quality electronics were produced.

Claude Shannon realized that the basic unit of information was a "bit."[7] This gave rise to the need to store information as bits of signals, or digital signals. Information needs to be transmitted in bits of signals in terms of 1s and 0s, called digital signals, rather than continuous signals called analog signals. In digital signals, the amplitudes of the signals are coded in binary numbers. Then, they are converted to precise digital signals. Hence, intuitively, one surmises that digital signals need more bandwidth to transmit a signal. The bandwidth is needed to ensure the precision in the signals. Parity check methods are used to ensure that the sequence of digital signals is precisely transmitted.

As we enlarged our knowledge to manipulate digital signals, then came the age of computer and information science.

INFORMATION AND COMMUNICATION TECHNOLOGY

Because its broad frequency range and validity, electromagnetic theory has been used ranging from classical electromagnetics to quantum electromagnetics. Electromagnetic sources from kilohertz to megahertz have been used since the very early days of radio wave engineering. They can carry voice signals as well as music and have greatly enriched the lives of humanity.

Another great gift of electromagnetics is its use in communication. Humans can now communicate with each other almost at the speed of light. Electromagnetics can carry a signal that goes around planet Earth about seven times per second. This is almost instantaneous in many applications. Because of this, collaborators can now discuss ideas as if they are next door to each other, greatly enriching international collaboration. This would not have been possible if Earth were large. Via the use of the internet, researchers and scholars can have video meetings for international synergy and interactions that change how we collaborate internationally. Some scholars tout this as "the Earth is flat" due to our proximity to each to other after the coming of the internet. For instance, manufacturing projects can now be designed in one part of the world and then almost instantly made in another part of the world.

NANOTECHNOLOGY AND LITHOGRAPHY

Nanotechnology, which has been an important driver in the growth of many modern technologies,[8] intrinsically capitalizes on the growth of nanolithography. Our hair on average is about 50 to 70 micrometers in diameter, but we can engineer a cube or a transistor that has dimension of a few nanometers. As a consequence, we can pack millions of transistors in a space as small as a strand of hair. Hence, memory has become very cheap because memory chips can now be made extremely small, amortizing their cost of production. This compounding effect of exponential growth in memory capacity has made memory chips almost free. Some companies such as Google give out email accounts for free.

COMPUTERS AND COMPUTATIONAL TECHNOLOGIES

Because of the growth of modern technologies, computers can now compute a lot faster. More memory and computational power can be packed into a smaller space. In the beginning, computers made of vacuum tubes, because of their sizes, could easily fill a room. But with the growth of nanotechnologies, now cell phones that people carry can have billions of transistors. That is almost as numerous as the number of neurons that God has given us in our brain. These computers can perform calculations at lightning speed. The growth of lightning speed for computations and algorithms has allowed the simulation of problems that once took years to now be done in a short period of time. One important consequence of this is the popular demand to replace real experiments in the lab with virtual prototyping and use digital twins to reduce design turnaround times. This has spurred the growth of computational electromagnetics in which Dan Jiao, Luis Gomez, and Thomas Roth, my colleagues at Purdue, are performing world-class research.

The rise of rapid computers has spurred another important growth: artificial intelligence (AI). Because computers can now compute a lot faster and memory is a lot cheaper, data-driven expert systems, the new AI, are in vogue. These systems can work a lot faster than the human mind and hence have great potential in regard to replacing the meaning of "work" in the modern world.

THE NEXT 50 YEARS

I have summarized what has happened approximately in the last 150 years. It is harder to project what is going to happen in the next 50 years. By studying the knowledge we have created in the last 150 years, we have many unsolved problems that we can embark on solving in the next 50 years. Many problems remain unsolved, and it will be foolish for me to conjecture what their solutions are. It will be more stimulating to list what these unsolved engineering problems are and challenge all to solve them:

1. Quantum technology and information science;
2. Biotechnology and bioengineering;
3. Global warming and environmental sustainability;
4. Eradicating corruption, human poverty, and inequality;[9] and
5. Eradicating human tribalism, racism, and the penchant for violence.[10]

The above are earthly problems affecting us. One can also think of heavenly problems such as space travel and the colonization of other planets and space. To me, the earthly problems are more important than heavenly problems!

1. *Quantum technology and information science.* We humans have a constant quest for technologies that helps us solve problems in exponential faster amounts of time. Quantum computers offer such possibilities. Quantum communications and quantum networks promise communication of exponentially large datasets in a finite time.
2. *Biotechnology.* There are many unsolved problems in biotechnology, such as why the human brain is so efficient in energy consumption compared to modern digital computers. The human brain consumes

about 25 watts of power, while a modern digital computer consumes thousands of watts of power to perform the same task.

3. *Carbon dioxide.* Due to negligence of the human species, we have spewed over 40 gigatons of CO_2 into our atmosphere.

ERADICATING POVERTY AND PROMOTING EQUALITY, PEACE, AND GLOBAL COLLABORATION

China, with accumulated wisdom over five millennium, has a tradition of managing large projects. The country has used its wisdom to lift 800 million people out of poverty. The human mind is one of the greatest natural resources and is a gift from God, just as is technology. Peoples of the world should collaborate to save our planet from global warming.

NOTES

1   B. J. Hunt, Oliver Heaviside: A first-rate oddity, *Physics Today*, 65(11) (2012), 48–54.

2   W. C. Chew, *Lectures in electromagnetic field theory*, Purdue University, Fall 2023, updated July 2, 2024, https://engineering.purdue.edu/wcchew/ece604f23/EMFTEDX070224.pdf.

3   *Gustav Kirchhoff*, Wikipedia, https://en.wikipedia.org/wiki/Gustav_Kirchhoff.

4   *Telegraphy*, Wikipedia, https://en.wikipedia.org/wiki/Telegraphy.

5   *Anomalous magnetic dipole moment*, Wikipedia, https://en.wikipedia.org/wiki/Anomalous_magnetic_dipole_moment.

6   Chew, *Lectures in electromagnetic field theory*.

7   *Claude Shannon*, Wikipedia, https://en.wikipedia.org/wiki/Claude_Shannon.

8   *Nanotechnology*, Wikipedia, https://en.wikipedia.org/wiki/Nanotechnology.

9   S. Chayes, *On corruption in America: And what is at stake*. Vintage, 2021.

10 J. S. Spong, *The fourth gospel: Tales of a Jewish mystic*. HarperCollins, 2013.

4

OUR ENGINEERED FUTURE

Written by **Ms. Rosa** and students in her fourth- and fifth-grade classes: James C., Nyle D., Brynn D., Barrett D., Quinn F., Maxwell H., Ava H., Karenza H., April J., Colt M., Carson M., Andrew N., Kylie R., Ava R., Sebastian S., Alexander S., Theodore S., Jackson S., Marcus S., Adalee S., Madeline S., Gabriel T., Henley V., Gavin W., Joshua W., and Monroe W.

Purdue engineering has made so many big accomplishments throughout its 150 years. Since 1882, Purdue has established 11 schools of engineering, including the School of Aeronautics and Astronautics, the School of Nuclear Engineering, and, most recently, the School of Agricultural and Biological Engineering. Purdue engineers have gone on to accomplish incredible things, such as Neil Armstrong, who was the first person to walk on the moon; Amelia Earhart, who worked at Purdue and later went on to be an extraordinary aviator; and Charles Alton Ellis, who helped design the beautiful Golden Gate Bridge. To us, engineering is all about solving problems to make the world better through science, creativity, and perseverance. Now, we invite you to go on a journey with us through our very own imagination to the future of engineering!

Let's face it, climate change is getting pretty bad. So, you might be wondering how we can solve this. First, let's cover the effects it has on animals, humans, and the earth around us. No one wants to go outside in December to play in the snow only to find that it's 80 degrees outside! Well, that is exactly what will happen if we don't do something about climate change. Even the Arctic could melt, and it's already starting to. Over the past 35 years we have lost about 28 trillion tons of ice, and more than 80% of natural disasters are caused by real-world aspects related to climate change. Climate change could be unstoppable by 2030. So, here's an engineered solution to this seemingly impossible crisis, one that we could use in 50 years, in the future!

Our idea to fix this crisis is to engineer a satellite, but not just any satellite. This satellite could go up to the ozone layer (the layer in our atmosphere causing most of global warming and climate change) and examine and study it so that we know more about this dangerous cause and know how to fix it. Some great causes of climate change are manufacturing, cutting down a lot of trees, and using a lot of water. Engineering could help us make factories more eco-friendly and use less resources that our Earth needs. Let's talk energy. Wind turbines are very helpful and highly eco-friendly. We could replace non–environmentally friendly generators with smaller but still very powerful wind turbines.

Another huge problem that humans are experiencing is hunger and thirst. One in 11 people, or roughly 733 million, experience hunger problems, and one billion people are affected by dirty water. You have probably heard about world hunger and the global water shortage, and if so then you have probably also heard that they are major problems that affects almost 10% of the humans on Earth. Welcome to the future! One way to solve our two shortages is to engineer a better way to power artificial intelligence. To power artificial intelligence you need to collect water, and collecting water takes away from the people in need. Another way to save and preserve water is to design wells so that they are more durable and collect more water. Maybe they even clean the water while it's being collected!

Engineering can help us power this new technology without hurting our resources. World hunger started in about 1972, and one reason it is happening is because people are wasting most of their food scraps. One way to solve world hunger is with agricultural engineering. We could figure out ways to make crops grow faster and make food stay fresh longer. These examples are only a few ways we can use engineering to solve world hunger and the water shortage.

Engineering can also inspire us to discover more about our universe. Our universe has been rapidly expanding, and thanks to many smart people at Purdue, our knowledge about space has expanded too. We have been exploring space for over 50 years. The Boilers to Mars program, one of the many programs at Purdue, will help take Purdue students to Mars and will help them develop rockets capable of long-distance space travel. Mars is a cold, rocky place. It is the fourth planet from the sun and is commonly known as the Red Planet. Many rovers have explored Mars, but we think that in the future of space engineering, *people* will head to Mars and explore it in person! Maybe humans can find water sources if they're able to stay on Mars longer. Can we create an oxygenated bubble that scientists can roam in without need for space suits? What about engineering a way to preserve seeds to transport to a new planet? Can we build a greenhouse on Mars to experiment with soil and growing plants?

Mars could even be a checkpoint. Once we're there, we could build telescopes and spacecrafts to get us even farther in space. Will we discover or create water or life on other distant planets? With more intelligent engineering, we can imagine auto space flight capabilities and more powerful telescopes.

We have learned all about engineers' past accomplishments, problems our world has had, and the possible solutions to these few problems, such as climate change. We've learned about how it affects people, animals, and our whole Earth. What about world hunger and the water shortage? This is what engineering is for, to come up with solutions that are best for our home planet. And space, how can we keep inspiring discovery? Our world has its problems, but we can change that with the power of collaboration, advancement, and imagination that engineering is all about. Think about all the solutions we have come up with. Could you think of better solutions? Become an engineer! Use *your* ideas to change the world and its future!

REFERENCES

Action Against Hunger. (n.d.). *World hunger facts.* https://www.actionagainsthunger.org.uk/why-hunger/world-hunger-facts

Bater, Simon. (2014, March 26). *Big facts: Climate impacts on people.* Climate Change, Agriculture and Food Security, March 26. https://ccafs.cgiar.org/news/big-facts-climate-impacts-people

Children International. (n.d.) *Global poverty and hunger: World poverty facts & Global Hunger Statistics.* https://www.children.org/global-poverty/global-poverty-facts/facts-about-world-poverty-and-hunger

NASA. (2013, October 23). *The IPCC's four key findings.* https://science.nasa.gov/resource/graphic-the-ipccs-four-key-findings

NASA. (2025, July 1). *NASA's impact.* https://www.nasa.gov/nasa-impact

National Security Technology Accelerator. (2024, September 23). *AI driven advancements in space.* https://nstxl.org/ai-driven-advancements-in-space

Purdue University. *Boilers to Mars.* https://boilerstomars.com/

Raman, A., & J. Small. (2025). Celebrating 150 years of Purdue engineering. Purdue University Press.

5

AN ENGINEER'S JOURNEY INTO THE FUTURE

KARA CUNZEMAN

Kara Cunzeman is the principal director of the Enterprise Lab (eLab) at the Aerospace Corporation. She earned her bachelor of science degree in multidisciplinary engineering, in 2008 and her master's of science degree in aeronautics and astronautics engineering in 2010, both from Purdue University.

COMMENCEMENT SPEECH AT PURDUE UNIVERSITY, CLASS OF 2075

The future belongs to those who believe in the beauty of their dreams.
—Eleanor Roosevelt

These words have guided me throughout my life and career. Today, as I stand before you—graduates of Purdue University's Engineering Class of 2075—I am reminded of the boundless potential you hold to shape the world of tomorrow. You are engineers, dreamers, builders of futures yet to be imagined. It is your courage, ingenuity, and curiosity that will define the next half century, just as it has shaped my journey.

Let me take a moment to share my story, not because it is extraordinary but because it is a testament to the power of engineering and how it can transform lives, solve humanity's greatest challenges, and open doors to the stars—literally!

MY JOURNEY INTO ENGINEERING

I grew up in a small town in Indiana not far from here. My earliest memory of engineering was hanging out with my dad, watching him explore the universe on our virtual reality headset in our living room, and asking questions to our favorite artificial intelligence (AI) assistant until it was stumped. My father had been a mechanical engineer in his time, and although his tools were simpler by today's standards, I was fascinated by his ability to build almost anything, in real life and in the digital world and by his insatiable curiosity. "Your world is going to look radically different than the one I grew up in, kiddo," he would often say to me. And indeed, a great jump in capacity across all facets of society were ignited by AI. "But," he told me, one thing will stay the same: the power of a single human to dream beyond themselves."

That always stuck with me, that recognition that I am unique because I dare to dream, to create, to think boldly. That I am human.

In school, I gravitated toward all things. I was a renaissance woman! I did love math and science, but I also loved the arts and creating. But it wasn't until my high school robotics team built a drone to help local farmers monitor their crops that I truly understood the power of engineering. Watching our creation take flight—seeing its impact on our community—lit a fire in me. I knew I wanted to be part of something bigger, to solve problems that mattered.

When I was accepted to Purdue University as an undergraduate in 2050, I felt like I had been handed the key to a new world. Little did I know just how far it would take me.

PURDUE UNIVERSITY: THE LAUNCHPAD TO MY DREAMS

Purdue has always been a cradle of innovation, a place where the impossible becomes possible. From the moment I stepped onto campus, I felt a sense of history, of walking in the footsteps of legends such as Neil Armstrong and Amelia Earhart. But I also felt the pull of the future, of becoming part of something greater.

As an undergraduate I studied living systems engineering, a field that had evolved by integrating biology and engineering through the design of living structures. I worked on projects involving smart bio farms—autonomous ecosystems that combined synthetic biology, algae-based energy systems, AI, and robotics to produce food with minimal environmental impact. These farms were part of a new wave of agricultural engineering, designed to feed a growing population while healing the planet.

But it wasn't just the work that inspired me—it was the people. My professors, mentors, and peers challenged me to think bigger, to collaborate across disciplines, to leverage cutting-edge AI systems, and to embrace failure as a stepping stone to success.

When the opportunity arose to pursue doctoral research on the moon, I didn't hesitate. The Commercial/Collegiate Lunar Research Initiative was a joint effort between Purdue and several other universities across the globe in partnership with a consortium of private space exploration companies, and it was humanity's first attempt to create a permanent research base beyond Earth. My focus? Developing construction materials from lunar regolith—moon dust—to build habitable structures that could withstand extreme conditions.

Living and working on the moon was both exhilarating and humbling. As I gazed at Earth from 238,900 miles away, I was reminded of the fragility and beauty of our home. It was a privilege to contribute to humanity's first steps toward becoming a multiplanetary species. But it was also a stark reminder of the challenges we face—both on Earth and beyond. I was also reminded of the gift and obligation I had as a human to continue to dare to dream. Those big dreams I had as a kid were why I was standing on lunar soil at that moment.

A WORLD READY FOR THE NEXT
GENERATION OF ENGINEERS

Class of 2075, the world you are stepping into is as full of promise as it is of challenges. Over the next 50 years, engineers will be called upon to solve problems that are as vast as they are complex. Let me give you a glimpse of what lies ahead.

AGRICULTURE

With Earth's population projected to reach over 11 billion by 2100, we must revolutionize food production. Engineers will design vertical farms that float in the atmosphere, harnessing solar energy to grow crops in arid regions. Synthetic biology will allow us to create nutrient-dense foods and supplements from carbon dioxide and water, with the potential to eliminate the need for traditional farming altogether.

ENERGY

The global transition to renewable energy is well under way, with the next frontier being in scaling our incredible advancements in fusion power over the last decade. We now are in a place to start scaling clean, limitless energy derived from the same process that powers the stars. Engineers of this generation will also develop energy storage systems capable of sustaining entire cities, ensuring a stable power supply even during extreme weather events.

MEDICINE

Advances in bioengineering will continue to redefine health care. Imagine neural interfaces that can repair spinal injuries and on-demand synthetic organs that are grown in bioreactors. Engineers will design personalized medical treatments using quantum computing and AI, making diseases such as cancer and Alzheimer's a thing of the past. Your generation will even be working on cutting-edge genetic modification techniques that will allow humans to thrive off-world!

CONSTRUCTION

Engineers will build cities that float above rising seas, structures that adapt to earthquakes in real time, and habitats on Mars and beyond. The materials of tomorrow—self-healing concrete, programmable metals, and bioplastics—will redefine what is possible.

SPACE EXPLORATION

The moon that I had the privilege to call home for a short time was only the beginning. By 2075, engineers will be looking to establish human presence on Mars, demonstrate scaled asteroid mining operations, and design interstellar autonomous systems. Your generation will write the next chapter of humanity's journey into the cosmos.

THE ROLE OF AI IN ENGINEERING

None of this will be possible without the transformative power of AI. AI has already revolutionized engineering education, allowing students to access personalized learning pathways, simulate complex systems, and collaborate with peers across the globe in virtual environments, all at remarkable speed. What took me eight years in school to achieve decades ago can be achieved in a student's first semester!

In the workplace, AI acts as both a partner and a tool. Imagine designing a new kind of skyscraper with the help of an AI assistant that can predict structural weaknesses, optimize materials, and generate blueprints in real time through novel approaches in design. Or imagine working on a global team where language barriers are eliminated by AI-driven translators, enabling seamless collaboration.

AI has also given rise to entirely new disciplines within engineering. At Purdue, students now pursue degrees in quantum computing engineering, neural interface design, biophilic digital architecture, and climate adaptation engineering. These fields didn't exist 50 years ago, but today they are at the forefront of innovation.

Yet, as powerful as AI is, the true power lies in human creativity, empathy, and vision. Engineers are not just problem solvers—they are storytellers, dreamers, and builders of a better world.

THE FUTURE IS CALLING

Class of 2075, you hold the torch now. You are the architects of the future, the ones who will take us further than we have ever gone before. But remember: with great power comes great responsibility. The challenges we face—global water and food scarcity, energy transition and sustainability, aging populations and health care, the leap to other planets, and the preservation of our planet—are immense. Yet, so too is your potential to overcome them.

As you leave this university and step into the world and perhaps for some of you to the stars, I urge you to dream big, to take risks, and to embrace the unknown. Don't be afraid to fail; every failure is a lesson, and every lesson brings us closer to success. Collaborate across disciplines. Seek out perspectives different from your own. Your humanity is your strength.

And most of all, never lose sight of the wonder that brought you here. Engineering is not just a profession—it is a calling. It is the belief that we can build a better world not just for ourselves but also for generations yet to come.

So, go forth, class of 2075, and create the future. Solve the unsolvable, dream the impossible, and reach for the stars. The world is waiting for your light.

Congratulations, engineers. Now, let's get to work!

This story was written by human-AI collaboration between futurist Kara Cunzeman and AI assistants provided by you.com.

6

THE NEXT 50 YEARS

JUAN M. DE BEDOUT

Juan M. de Bedout earned his bachelor's, master's, and doctorate degrees in mechanical engineering from Purdue University. He is the chief technology officer at RTX, where he is responsible for shaping the company's technology strategy and driving excellence in engineering.

I was born in 1972, just four years shy of the 100th anniversary of the school of engineering at Purdue. It was a momentous time, just a few years after the historic Apollo 11 moon landing, and people all over the world shared in the marvel of what that historic challenge had produced. It was a proud moment for my grandfather Lawrence Cargnino as well, a professor in Purdue's School of Aeronautics and Astronautics who had the fortune of having been a part of the development of several of the astronauts who had made the moment possible, including Neil Armstrong, Roger Chaffee, and Gus Grissom. Nobody in the School of Aeronautics and Astronautics or anywhere else in the world for that matter would have believed that this chapter of human exploration would close so abruptly, with humanity's last walk on the surface of the moon about to take place on December 14, 1972, by another Purdue graduate, Eugene Cernan.

The 100 years preceding the moon landings were a remarkable demonstration of human innovation and perseverance. Having just emerged from the industrial revolution, many people's existence in the 1870s was still one of drudgery and daily toil to put food on the table. What came next must have been dizzying to a person at the time, with, to name just a few examples, the invention of the electric generator by Werner von Siemens in

1867, building on the earlier concepts of Michael Faraday and the subsequent commercialization of DC electric power by Thomas Edison in the early 1880s and of AC electric power by George Westinghouse a few years later; the invention of the incandescent light bulb in 1879 and its commercial introduction the following year; the invention of the automobile in 1886 and its mass production starting in 1908; the invention of the telephone in 1876 followed by transcontinental cables in 1915; the development of modern vaccines for pertussis in the 1920s and the polio vaccine in 1955; and the invention of flight by the Wright brothers in 1903, followed by the subsequent use of airplanes in World War I as early as 1914. Theoretical physics grew by leaps and bounds as well, with Albert Einstein's formulation of special relativity in 1905 followed by general relativity in 1915 and with the work of Max Planck and Niels Bohr leading to the description of electron energy level quantization in 1913, laying the early foundations for modern quantum mechanics. The pace of innovation continued with the emergence of the great power competition between the United States and the Soviet Union, with the development of jet engines that enabled supersonic flight and rocket engines that enabled hypersonic flight, access to space, and our historical trips to the moon. The period also saw the development of nuclear fission and fusion that enabled the dual possibilities to energize the world or destroy it and the development of semiconductors that led to the first transistor in 1947 and then to integrated circuits in 1958. Progress was fast across a diverse array of fields, building from modest prior knowledge and driven by a tenacious curiosity to explore the boundaries of science and technology. While costly to develop, the aggregate return on these technology investments, to both the nation and the world, are incalculable.

The 50 or so years after the moon landings, bringing us to today, have also seen a fast pace of technological development although of a very different nature. What was once characterized by a staggering breadth of novel exploration across numerous undeveloped fields gave way to the steady refinement of technologies along proven paths in a search for how to continuously improve performance, utility, and cost. Transportation became safer and more fuel efficient, communication became wireless and digital, and navigation moved from maps to Global Positioning

System solutions. The first steps taken to sequence DNA in the 1970s led to low-cost sequencing and targeted gene editing capabilities in the 2000s, creating the opportunity to radically improve health care outcomes for everyone. The creation of ARPANET in 1969 led to the modern internet, which has made a wealth of human knowledge available digitally. And the recent evolution of large language model artificial intelligence (AI), enabled by cost-effective, high-performance computing, has made the possibility of scanning, summarizing, and interpreting that accumulated knowledge accessible to anyone with a smart phone anywhere in the world. Life has clearly become easier and more comfortable for people everywhere, and the stage has been set for this new accessibility of knowledge to fuel innovation by the broadest population to ever engage in contributing their new ideas to the world.

But we lost something in the transition between these two periods. Commercial flight has not become faster, cars remain confined to roads where congestion can slow vehicles to horse speeds, nuclear energy has been largely sidelined, access to space is limited to few, and we have not returned to the moon, let alone established a plan to put a person on Mars, to name a few examples. While it is true that scientists, engineers, and entrepreneurs are currently working on all these areas, they would now be mature had the momentum they carried in the early 1970s been maintained. The wind-down of the global power competition, in combination with a more globalized economy hyperdriven by consumerism, has led to a different set of national priorities where ambitions for technological leadership are often missing or have been incapable of surviving across the government policies of successive administrations, left instead to the passion of emerging entrepreneurs. While neither good nor bad, it is clear that the technology development choices we have made over the last 50 years have put us on a very different path from the one that was once ahead of us.

In this time frame, we also saw a decline in US manufacturing capability, and today's industrial supply chains are highly dependent on a broad mix of international sources, with a dynamic geopolitical environment posing risks that carry far-reaching economic and national security implications. As we are seeing, these tensions can impact what have been finely tuned global networks that until recently were driven primarily by market forces.

I am optimistic that our next 50 years will strike a different, more hopeful balance. We must dare to dream big as engineers and as Americans. National priorities targeting technology leadership in space, aviation, semiconductors, energy, advanced manufacturing, AI, genetic engineering, and quantum technologies should be the foundation of a strategy to revitalize the country's economy and bolster its readiness for the changing geopolitical dynamics taking shape in the world. Road maps charting the path to excellence in these areas will be essential to help influence our government leaders in setting national policy, and Purdue can play an important role in framing them. Key opportunities that our alma mater can help with include the following:

- Framing a vision for the next chapter of space exploration. Purdue's Cislunar Space initiative shares a bold strategy for launching a near-Earth space economy that can serve as the staging ground for a broader exploration of the solar system.

- Recapturing global leadership in both air-breathing and rocket-based hypersonic propulsion to enable reusable single-stage access to space and future high-speed transportation. Purdue's modernized Zucrow labs will be a national asset in this pursuit.

- Charting a path to US technology and manufacturing leadership in advanced semiconductors, photonics, and chip packaging technologies to ensure that the most sophisticated capabilities are available to defense and commercial applications. Purdue's investments in the Birck Nanotechnology Center and the innovation ecosystem it is creating through industrial partnerships provide a firm foundation to lead the charge.

- Leading the world in the development and integration of cost-effective clean and renewable energy, including modern nuclear fission and fusion. Purdue's LEAPS (Leading Energy-Transition Advances and Pathways to Sustainability) initiative brings thought leadership to growing wind and solar energy resources in the national grid interconnections. In the area of nuclear energy, key active research includes the university's collaboration with Duke Energy in exploring the use of small modular nuclear reactors and research toward developing advanced materials for fusion reactors at the Center for Materials Under Extreme Environments.

- Developing a plan to rebuild our national manufacturing competence with next-generation technologies, including advanced materials processing and production, additive manufacturing, and Industry 4.0 technologies that unlock productivity and efficiency, mitigating the foreign labor cost advantages that have driven the current paradigm. Purdue's Excellence in Manufacturing and Operations initiative is a superb program for helping bring thought leadership in this area.
- Shaping how AI will be integrated into the next generation of products to deliver transformative capabilities as well as how it will be integrated into the next generation of engineering design tools, raising the design proficiency of engineers throughout every industry and accelerating the speed and richness of product development. Purdue's Institute for Physical Artificial Intelligence and the Institute for Control, Optimization and Networks are well positioned to influence the national strategy for AI.
- Introducing novel tools and analytical capabilities in genetic engineering to unlock novel health care treatments and life-enhancing therapies for people around the world. Purdue is already making great strides in this field, with new gene editing technologies such as Natronobacterium gregoryi Argonaute that promise to surpass the capabilities of the industry's CRISPR-Cas9 gene editing workhorse that has driven so much progress in the last decade.
- Ensuring leadership in the emerging quantum computing, sensing, and communications fields, which will revolutionize applications ranging from advanced aerothermal design for aerospace systems to secure communications for commercial and defense needs. Again, Purdue has a strong team in place, with the Purdue Quantum Science and Engineering Institute well positioned to lead the way.

Progress in these areas will assuredly take place, and the open question now is who will lead and reap the rewards. Establishing a vision is a critical first step. Translating it into action will be the longer, harder task that will require focus, perseverance, and hope. But if we are to truly believe that our better days are ahead of us, we must take upon ourselves the task to see this through. When we succeed, we will see a time of change unlike anything those of us alive today have witnessed before. My grandfather would hardly recognize the world coming out of this next period except for the frenetic pace of innovation that will hopefully be reminiscent of the long-gone days of his youth.

7

ENGINEERING THE 2075 FUTURE OF AIR TRANSPORTATION

Faster and Cleaner

DANIEL DELAURENTIS

Daniel DeLaurentis, PhD, is the vice president of Discovery Park District Institutes and the Bruce Reese Professor of Aeronautics and Astronautics at Purdue University. He is a fellow of the American Institute of Aeronautics and Astronautics and the International Council on Systems Engineering.

THE VISION: GETTING FROM POINT A TO POINT B AS FAST AS POSSIBLE

My vision of air transportation 50 years from now is born of experiences that began 30 years ago. In 1995, I began graduate research at Georgia Tech funded by the National Aeronautics and Space Administration (NASA) in pursuit of our nation's goal of achieving an economically viable, operationally feasible supersonic commercial transport aircraft. After all, who wouldn't want to transit the Pacific Ocean in three hours rather than

10 hours? This truly nationwide effort spanning government (NASA), industry, and academia concluded that the design space of viable aircraft configurations was empty! The full range of integrated technologies, operational procedures, and business models required to make this work were simply not invented yet (and even today are only emerging in small and mostly isolated instances). Yet many people (then and now) remain captivated by such flight, and studies have assessed that passenger demand for safe, affordable supersonic air travel remains strong. But I wondered, would we get there by 2050 or even 2075?

A decade later in early 2005 I embarked on my faculty career at Purdue University's storied School of Aeronautics and Astronautics. My first sponsored research grant, again from NASA, was to develop design tools able to generate what NASA termed "revolutionary system concepts for aeronautics" (RSCA). My proposed RSCA test case was a robust, scalable multimodal transportation system of systems tailored for personalized air travel and designed to maximize the speed for travelers to get from point A to point B, from doorstep to destination. Since Purdue hired me under its College of Engineering system of systems signature area, I was thrilled to receive this endorsement from NASA to put these new theories and methods to work. To top it off, just as my project was commencing, the image on the cover page of the November–December 2004 issue of the bimonthly research magazine *TR News*, published by the Transportation Research Board (a unit of the National Academies of Science, Engineering and Medicine), captured the essence of my RSCA idea: a transportation future that would fulfill dreams of seamless, rapid, multimodal transport optimized for each individual yet be efficient in the aggregate!

Our research discovered new ways to pose the problem, model the system of system dynamics, and derive the interrelated technological advancements and operational paradigms required for such speed in getting from point A to point B in a local/regional setting. Yet once again, these required innovations were just not ready for prime time. But I wondered, would we get there by 2050 or even 2075?

Fast-forward another 20 years to our present day, 2025, and once again the dream of fast and efficient local, regional, and global aircraft-based multimodal transportation seems tantalizingly close to reality. Regarding supersonics, researchers at Purdue and elsewhere have conducted fresh studies on the market potential, fleet composition, and technological requirements for rejuvenated supersonic air service. NASA's X-59 quiet supersonic research aircraft is about to fly and demonstrate low boom technology that could sufficiently mitigate sonic boom such that it would be acceptable for supersonic overland flight. On the urban and regional front, an entire new industry has emerged around the concept of advanced air mobility. This includes missions such as urban air mobility, which would carry passengers in four- to eight-seat electric vertical takeoff and landing aircraft on commuter trips around the largest metro areas as well as middle-mile and last-mile cargo delivery served by small and medium-sized electric autonomous air vehicles. Again, researchers in Purdue's School of Aeronautics and Astronautics are among the nation's leaders in this exciting research area that some have dubbed "Uber in the sky." Further, hundreds of companies have raised billions of dollars and are advancing the aircraft, vertiport, and air traffic management concepts that will form the heart of the services. We are getting there, but will we arrive by 2050 or even 2075?

THE METHOD

Modeling and evaluating the projected behavior of the existing air transportation system, let alone possible future ones, is overwhelmingly difficult due especially to the extreme number and heterogeneity of independent systems, the distributed nature of these systems, and the presence of deep uncertainty in understanding their coevolution. In light of these properties, such complex systems are best described as a system of systems, a collection of diverse things that evolve over time, organized at multiple levels, to achieve a range of (likely) conflicting objectives but never quite behaving as planned. I have based my entire academic career on this foundational idea: we can indeed systematically integrate disparate and independent technologies, operational models, and business models that collectively adapt and evolve, ever generating more effective new capabilities and services at scale if they are properly modeled as systems of systems.

THE OUTLOOK

Based on these past experiences and the present state of affairs and with confidence in our collective system-of-systems research and innovation acumen (at Purdue and beyond), I foresee three key hallmarks that will bring us, finally, to our desired destination, that is, a truly effective multimodal, aircraft-based transportation system that allows people to get from point A to point B with speed, safety, and ease:

- *Diversity (of aircraft types).* Finally, we will have broken free from the tube-and-wings architecture for aircraft design and enjoy the benefits of a diversity of aircraft types and sizes optimized for particular missions and using a variety of fuels, all environmentally and economically sustainable in nature including electric, hybrid-electric, and hydrogen powered.
- *Autonomy (everywhere).* Far more than just the piloting of the aircraft will be automated. Provably safe automated systems will execute the air traffic management that maintains and even surpasses today's amazing aviation safety record. Artificial intelligence systems will constantly monitor environmental conditions and automatically mold the flight operations to avoid or mitigate bottlenecks, delays, and disruptions.
- *Seamlessness (for everyone).* The entire doorstop-to-destination passenger trip is monitored and managed by autonomous software agents that ensure maximal service experiences. Mode changes will become seamless and almost imperceptible to the passenger, with a dramatic increase of total trip efficiency at a fraction of the carbon emissions seen today.

It is my fervent hope that we will have arrived long before 2075!

8

ENGINEERING 50 YEARS FROM TODAY

KENZIE

Kenzie is a fifth grader in Mrs. Lehman's class.

Kenzie Polk - Mrs. Lehman's 5th grade

Engineering 50 years from Today

In fifty years I think that electrical engineering will be way more advanced than it is now. I also think there will be a broad field of technology with devices, such as holograms for communication instead of facetime. In my opinion the electrical field of engineering is going to take off. With todays generations being very connected to devices I think there will be more electrical development with technology in fifty years. It's also possible that in fifty years electrical engineering will take off because of our generations following in the footsteps of past engineers making a very clear path for our generation.

In fifty years I think the biomedical department will be the most advanced as its ever been. I think there will be cures for cancer, and many diseases that they haven't found a cure for yet. Another thing I think biomedical engineering will create medicine to help keep us healthier. I also think biomedical engineering will come out with special types of food that have protein but still taste like the regular unhealthy foods. I also think they will come out with drinks to help with asthma and Pheumonia.

Kenzie Polk — Mrs. Lehman's 5th grade

Engineering 50 years from Today

In fifty years I think aerospace engineering will be very advanced. I think they will find a way to make space and air travel faster. I'm sure they will find a way to make traveling in aircraft and spacecraft safer for humans. In fifty years I think they will make the inside of planes and spaceships more comfortable with more food and drink options. Another thing I'm sure they will do in fifty years is make better designs for aircraft and spacecraft with many more advantages.

In fifty years I think there will be many opportunities to improve our lives. In fifty years I am sure will be a great sight to see. I'm sure without a doubt that there will be many more amazing inventions coming in our future. I just want to say that nothing would be possible without the engineers before us that have paved a path in the engineering world helping us understand more from their work. I am personally looking forward to the next fifty years to see all the amazing things today's generation will have created.

9

INDUSTRIAL ENGINEERS ARE POISED TO TRANSFORM AMERICA'S ECONOMY

SUE ELLSPERMANN

Dr. Sue Ellspermann began her career in industrial engineering and is the president of Ivy Tech Community College. She previously served as Indiana's lieutenant governor and as a state representative.

THE EVOLVING INDUSTRIAL ENGINEERING FIELD

As a freshman engineering student at Purdue University, I had no idea what kind of engineer I wanted to be. I was, however, proud to be part of Purdue's burgeoning Women in Engineering cohort, representing 10% of our freshman class. While I knew that guys from my high school were going to major in engineering, I didn't know a single engineer and had been dissuaded from the field by my high school counselor. Thank goodness for

Engineering 100, where each week we heard from practicing engineers in the various disciplines. When I learned what industrial engineers (IE) did—productivity improvement, human factors, facility planning, simulation, and more—I was hooked.

When I graduated in 1982, most IEs were employed in manufacturing. I had been an IE co-op student at AC Spark Plug, a division of General Motors located in Flint, Michigan, throughout my education. There, I worked in production planning, human factors, facilities planning, and packaging. Michelin Tire Corporation, my first employer after college, used IEs for time studies, pace rating, and facilities design. Frito-Lay, my third employer, used IEs for productivity improvement in a much broader definition: process improvement, facilities improvement, automation, logistics, research and development, and even sales operations.

By the late 1980s and the 1990s, IEs were increasingly recruited into health care to design emergency rooms and improve workflow of complex systems, eventually driving toward electronic medical records and further digitization and automation. And there were a lucky few IEs employed by Disney to improve flow, reduce customer wait times, and improve the customer experience in its parks and resorts. Who wouldn't want to have that job? True imagineering.

After stints in production, shipping, and plant engineering at Frito-Lay, I joined the corporate industrial engineering team, which also oversaw the hiring of nearly 100 IEs per year. These engineers were deployed to 40 plants, sales operations, logistics, and headquarters focused intensely on productivity improvement. I became engaged with and helped lead Frito-Lay's methods improvement program called OFFSET, named after the need to offset the impact of rampant inflation experienced in the 1980s. It was understood that snack foods were not a necessity, and if not controlled, these products would outprice the market. An aggressive goal was set to negate the effect of inflation: saving $500 million in five years using Simplex creative problem-solving as a tool for methods improvement and cross-functional innovation.

By 1986, I was an external consultant in Simplex and pursued my master's and doctorate degrees at the University of Louisville Speed School Department of Industrial Engineering. There, I explored the application of unstructured problem-solving to better understand solving "messy" problems. My PhD adviser was, of course, a PhD from Purdue, Dr. Gerald Evans. That research was recognized in 2007 as a Top 50 management research article by *Emerald Management Review*.

AN EVERGREEN SKILL SET

After entering higher education as the founder of the University of Southern Indiana Center for Applied Research and Economic Development and completing a term as an Indiana state representative, I was elected as Indiana's lieutenant governor. Agency budgets had already been slashed in the aftermath of the Great Recession, and new moneys were not to be found with Governor Mike Pence, who further cut taxes. The answer was to do more with less. That meant applying industrial engineering skills and tools to the six state agencies I oversaw. We implemented lean principles, Simplex creative problem-solving, and shared services across the agencies. We increased collaboration and began to innovate across agencies. We learned to not only survive but also thrive with fewer resources.

In 2016 I became the president of Ivy Tech Community College, the nation's largest singly-accredited statewide community college. Community colleges had been losing enrollment since the Great Recession, and Ivy Tech was no different. In fact, several of our campuses were nearly bankrupt when I took the helm. After touring the 6.5 million square feet of our large state campus system in my first month, I set a goal to reduce 1 million square feet of our footprint in an effort to reduce our operating expenses. We restructured the college to 19 campuses to ensure that our academic and skills training offerings were fully aligned to the diverse needs of Indiana's employers and communities. We hubbed more than a dozen back-office tasks that can be done at one location, often at one of our 19 campuses, for the entire college. We created accountability to a single strategic plan and key performance metrics. Further, without major reductions in force, we right-sized faculty and staff in proportion to current enrollment, which is once again growing healthily, as we served more than

200,000 students in the 2024–2025 academic year. We developed IvyOn-line, which consolidates online courses into one entity taught by our faculty while optimizing course sections at 25–30 students each, saving the institution nearly $2 million per year. Recognizing that our students also needed help, we introduced Ivy+ Textbooks to reduce book costs to less than $20 per credit hour, leveraging our negotiating power to save students more than $65 million to date. Further, when adjusting for inflation, students pay less in tuition today than they did a decade ago. These and many other operational innovations have brought us back to strong financial health and bond rating upgrades—the highest in the college's history, demonstrating our financial strength and lowering the cost of capital.

Industrial engineering teaches optimization, process improvement, and systems thinking. The latter is often understated. As the nation's largest singly-accredited statewide comprehensive community college system, Ivy Tech Community College acts as one college, with 19 campuses and more than 40 locations. Our more than 70 programs are offered on multiple campuses using the same curricula and learning outcomes. As with Ivy+ Textbooks, IvyOnline, and comprehensive programs, our students enjoy some of the strongest transfer agreements in the nation because of our scale. We have become a competitive advantage to the State of Indiana, our major employers, and the Indiana Economic Development Corporation by leveraging our ability to fill robust workforce pipelines as one institution with 19 campuses. For instance, Ivy Tech hosts the nation's largest associate in nursing degree program and graduates nearly 1,500 nurses each year. In fact, one out of every three registered nurses in Indiana is an Ivy Tech graduate. Further, half of all manufacturing credentials earned in Indiana each year are through Ivy Tech, which provides outsize support for Indiana's thousands of manufacturing enterprises and over half a million Hoosiers employed in Indiana's largest economic sector.

Like many of my engineering classmates, I haven't had "engineer" in my title for more than 30 years. There is no doubt, however, that our Purdue industrial engineering education has prepared us for executive leadership in nearly every industry including government and education, where efficiency, productivity, and innovation are required to thrive.

THE FUTURE OF IES TO AMERICA'S ECONOMY

The IE of today is well versed in the Internet of Things, data analytics, artificial intelligence, and the advent of quantum computing. The optimizations and simulations that were not possible decades ago will be at the fingertips of all IEs and operations analysts. Complex systems will be easily modeled and optimized. Like robots, GPS, and ChatGPT, these new tools will make human work more productive and safe and make lives easier. With this productivity will come higher expectations of humans to interface with automation, digitization, and augmented intelligence. This will lead to significant workforce demand changes that are difficult to project with accuracy. We have already seen automation reduce labor-intensive, low-skill jobs, and it is now very likely that we will see reductions in professional roles such as accounting, business, logistics, marketing, facilities, design, and legal operations. Some contraction will be welcomed as the labor force of the future begins to shrink, with baby boomers retiring and fewer young adults entering the workforce. Many workers will require upskilling or reskilling. In fact, TEConomy estimates that 82,000 incumbent Hoosier workers per year will require training to upskill or reskill over the next decade.

My optimism and deep hope are that IEs will ensure that humans do what humans do best while growing American prosperity. We can design human work that is augmented by artificial intelligence, automation, and digitization while ensuring that humans manage machines and data rather than serving them. We can ensure that humans use our unique skills—such as caring for one another, empathy, unstructured problem-solving, and creativity—to drive progress. I am even hopeful that IEs will create robots to assist humans with differing abilities and physical challenges—such as artificial limbs, implantable medical devices, and cognitive aids—that will enable many more people to participate in meaningful work.

Few professions prepare the changemakers of tomorrow with more evergreen change-making skills. IEs are taught to be problem finders—that is, to see opportunities for improving every thing, every process, and every system. Few professions push harder to find better ways of leveraging technology. And while systems thinking is not new, teaching system of systems thinking is critically important for the ever-increasing complexity of organizations and the world ahead. Analytical tools will be important, but these change-making thinking skills will be timeless.

Fifty years is very likely beyond my glidepath. Thus, my aspiration is that Purdue industrial engineering 50 years from now, when my great-granddaughters and great-grandsons attend, will seek to balance human, economic, and sustainability factors to build an ever more peaceful, healthy, interconnected, resilient, inclusive, and prosperous world.

The Purdue School of Industrial Engineering—with one of America's perennial largest and highest-ranked programs—has left a legacy of impact on our state and nation and has the opportunity to grow that legacy by attracting and providing a strong, diverse talent pipeline of IEs to serve Indiana, our nation, and the world in all industries, including government, nongovernmental organizations, nonprofits, and faith-based enterprises. I am betting on the Purdue School of Industrial Engineering to ensure that its graduates have the evergreen skills and tools to meet the challenges of the future and to serve with excellence.

A SALUTE TO ONE OF THE GREATEST

I would like to end with a tribute to a special Purdue IE. There was no one more beloved in the Purdue School of Industrial Engineering than Professor James Barany. Every student who had him understood that he loved us, was our biggest supporter, and expected us to make a difference with our careers. I was one of thousands of IEs who were privileged to know Dr. Barany as an instructor while at Purdue and a mentor for years beyond. Dr. Barany died in 2011 after more than 50 years as a Purdue School of Industrial Engineering faculty member.

#BaranyBoilermakerForever

10

THE FUTURE ENGINEER

Navigating the Impact of Artificial Intelligence on Design and Innovation

STEVEN GATTMAN

Steven Gattman, an inaugural recipient of Purdue University's 38 by 38 award, is a mechanical engineer in the defense industry with experience leading technical execution of programs ranging from $5 million to $1 billion.

L ess than 50 years ago, engineers relied on pencil drawings and manual design methods to create parts, a practice that has undergone significant transformation over the years. Today, advanced technologies enable engineers to model and visualize parts like never before, forcing a major shift in engineering processes. In the seven years since my entry into the defense industry, there has been a noticeable change as the sector increasingly adopts model-based systems engineering and, for example, virtual reality tools, reducing the reliance on time-intensive tasks such as generating part drawings and iterating designs through multiple prototypes. While previous technological advancements have already reshaped these workflows, the introduction of artificial intelligence (AI) holds the potential to not only enhance how tasks are performed but also transform how problems are approached and conceptualized. In the future, engineers will evolve from traditional designers to problem framers or product architects, tasked with defining the scope of AI-generated designs. This transition will demand a broader skill set and a more comprehensive understanding of the requirements of the desired solution while also requiring a keen awareness of the impact of using these tools. As AI continues to evolve in the engineering design space, it will provide the tools for anyone to create innovative and unique solutions to problems, fostering creativity in ways we've never seen before.

AI is here, and it's not going away anytime soon. While the general public primarily leverages AI for everyday tasks and quick information retrieval, industries are harnessing its potential on a much larger scale. Businesses are integrating AI to automate and optimize operations as well as streamline complex processes through the advancement of generative AI. A great example of this can be seen when looking at how the evolution of AI in engineering design is narrowing the significant skill gap that currently exists between the different generations of engineers. A skilled engineering designer can create a functional part in 25% of the time it takes an average designer, and the design process can be over 10 times faster than that of a new designer. This efficiency is crucial for meeting tight project deadlines, but skilled designers are scarce, and most programs cannot afford to keep highly paid experts on staff for every task. This skill gap is gradually closing as the design process and tools have evolved.

Design software, such as Siemens NX, has become increasingly AI-driven, enabling such software to optimize parts and make generative modeling decisions autonomously. Rather than relying on the expertise of seasoned designers to create complex models, advancements in software now enable less experienced engineers to perform similar tasks in comparable time frames, potentially saving companies millions in development costs each year. Looking ahead, the rise of machine learning will further transform the design process. Over the next 50 years, engineers will become increasingly dependent on AI-driven design software that requires minimal manual input. This software will instantly account for key constraints—such as manufacturability, reliability, and compliance: during the initial design phase, eliminating the need to revisit these factors in later iterations. Additionally, analyses such as structural and thermal evaluations will no longer be separate steps. Instead, real-time feedback will be integrated directly into the design process, allowing for rapid optimization in days rather than weeks or months.

While AI-driven design tools help narrow the skill gap between engineers, they also introduce significant concerns. One major risk is over-reliance on these tools, leading engineers to design parts as a "black box" without fully understanding the underlying processes and limitations of the AI systems. A clear example of this is the use of synchronous modeling tools in Siemens NX.

Traditionally, when a part change was needed—such as adjusting the size or location of a hole—a designer would navigate the part history, locate where the feature was originally created, and update it accordingly. This method allowed designers to follow best modeling practices and predict how changes would impact subsequent features. However, synchronous modeling enables direct manipulation of geometry, like pushing, pulling, or adjusting faces and shapes, without delving into the feature-based history. While this approach makes it easier to modify complex models quickly, especially for those unfamiliar with the design, it also introduces risks.

If used incorrectly, synchronous modeling can compromise the integrity of a part's design. Once introduced into a model, particularly when the engineer doesn't fully grasp the software's assumptions, future changes to features in the model history can have unpredictable consequences, potentially breaking features upon future updates. For instance, an engineer might take a shortcut by shifting a few faces using synchronous modeling without considering the original design intent. Later if another engineer attempts to modify the original feature in the model history, it becomes impossible to predict how the synchronous modeling adjustments will react. Faces and holes might shift unexpectedly, and these changes could go unnoticed by someone unfamiliar with the part.

This example highlights why maintaining a deep understanding of AI-driven tools is essential as they become more prevalent in the industry. Simply using technology to complete a task is not enough. We must understand both what AI is doing and how it's doing it to prevent designs from becoming an unpredictable "black box."

As AI changes the tools engineers use in their day-to-day jobs, the capabilities of AI will not only reduce the skill gap inherent to engineers of different experience levels but also shift the focus from engineering specialties into more of a technical generalist focus. Traditionally, design optimization has required collaboration among specialists: a designer creates the part and collaborates with specialists such as manufacturing engineers for manufacturability feedback and systems engineers for requirements compliance analysis. With AI eventually being able to fill these different roles by taking things such as manufacturability and performance requirements into account when generating an initial design, the role of an engineer will shift toward becoming more of a problem framer and a design modifier. Over time, the need for specialized engineering roles, such as those focused on structural or thermal analysis, will diminish as AI systems can be trained to handle these tasks with simple inputs from someone without a full technical background in those areas. In this new role, engineers will serve as mediators between human needs, AI outputs, and societal impacts, ensuring that AI-generated designs align with real-world requirements. In the near future, engineers will need to start adopting a more generalized, interdisciplinary skill set, becoming technical generalists with a broader understanding of various engineering domains.

Engineering institutions such as Purdue University already teach undergraduate engineers how to break down and solve complex problems rather than focus on deep-diving into specialized engineering areas, so this shift may have little impact on curriculum at the undergraduate level. Greater impacts will likely be felt at the graduate level, reshaping advanced coursework as industry demand shifts away from specialization. This evolution may place greater emphasis on engineering ethics, reflecting its growing importance in the field. What distinguishes human engineers from AI is our ability to consider societal and ethical values in decision-making, and as AI reshapes industries, engineers will be called upon to address the ethical implications of these technologies and shape policy around the use of AI.

Engineering is going to change a lot over the next 50 years. Traditional design life-cycle processes are going to be turned on their head, and the roles of engineers are going to change significantly, but the integration of

AI in engineering tools will open up new possibilities for creativity in solving societal problems beyond traditional industry applications. Imagine a world where you could simply describe your needs to an AI designer, automatically send the output to a 3D printer, and have a fully functional object that meets your needs within a few hours. The days of hobbyists requiring experience with 3D modeling software such as TinkerCAD will be a thing of the past, allowing anyone to create custom parts tailored to their needs. As these innovations unfold, AI has the potential to dramatically change how we approach design both inside and outside of industry, making technology more accessible and fostering creativity in ways that were once unimaginable. As long as we embrace the change and continue finding new and exciting ways to utilize AI for good, we have an exciting 50 years ahead of us.

11

FROM HERACLITUS TO HYPERSONICS

REED GEIGER

Reed Geiger is a PhD student in Purdue University's Aerospace Engineering program. He grew up in Churubusco, Indiana, and received his bachelor's and master's degrees in aerospace engineering from Purdue.

> *Thermodynamics owes more to the steam engine than the*
> *steam engine to thermodynamics.*
> —Lawrence Joseph Henderson

The ancient Greek philosopher Heraclitus believed that the fundamental essence of reality was fire. While other philosophers postulated that the fabric of reality was water or air, Heraclitus's tapestry was woven from a substance of constant flux and energy whose motion and heat imbued all reality with life. Six hundred years after the life of Heraclitus, another Greek, Hero of Alexandria, described a machine called the aeolipile that used steam generated by boiling water to impart motion on a vessel placed over an open flame. This was the first recorded demonstration of the conversion of chemical energy to kinetic energy, the same transformation that powers worldwide transportation on land, air, and sea and generates power for electric grids around the world. Ancient curiosity and intuition created the aeolipile, and the concept lay dormant for centuries before it was worked and improved by rigorous engineering principles into the world-powering technology that it is today. As long as humanity is driven to create and understand, the dual forces of inspiration and analysis will continue to propel us to new thresholds of technological achievement and mastery.

In parallel with Heraclitus's framework of the fire as the building block of the cosmos, the fundamental essence of today's world is energy. The efficient extraction, storage, and use of energy in its myriad forms has propelled the world through innovations in every discipline of technology at breakneck speed over the past 150 years. Energy is not only the chemical bonds in a drop of oil or the momentum of rushing water; it is also the calories in a sheath of wheat and the focused effort of a worker laboring at a task. Ever in flux, like Heraclitus's flame, forms of energy and how humans have utilized it change throughout history, with the only constant being that by the combination of raw inventiveness and disciplined analysis we arrive at more efficient and more productive utilizations that power the next leap in technology. And so the cycle continues.

Many of the fundamental challenges humanity will need to solve throughout the remainder of this century are rooted in energy. In the public consciousness, chief among these may be the decarbonization of the energy and transportation sectors in order to combat climate change. The difficulty of this problem not only lies in the harnessing of new sources of renewable energy and the development of more efficient processes for existing energy sources; it also entails building an infrastructure for these systems in a way that does not interfere with the environment in a new but equally damaging way. Additionally, food security will continue to be an ongoing challenge due to the combination of anticipated global population growth until the middle of the century and the encroachment of the industry and housing necessary to support this growing population on arable land. By casting food as an energy input to the human body, the challenge becomes that of maximizing energy output from agriculture while minimizing the land and resources necessary to produce said energy.

Engineering is equipped to solve these grand challenges by harnessing energy into productive forms, and this harnessing of energy creates tools that unlock new methods of utilizing energy that repeats the cycle indefinitely. Doing so allows us to transform the world in unpredictable ways. The creation of the modern steam engine in the late 1700s allowed trains to not only transform expeditions to the far-flung territories of a country into single-day trips but also serve as the heartbeat of the factories that milled out mass-produced items, cheap and accessible enough for any individual to purchase. Yet at the same time, this new conversion of energy also began to fundamentally change our environment in ways not comprehended until over a century later. As the steam engine became ubiquitous over the course of the 19th century, the natural drive to innovate and the desire to do away with the large boilers and water tanks needed for steam engines led to the development of the internal combustion engine. The continual improvement of the weight and efficiency of internal combustion engines over several decades was one of the final steps required to solve the problem of human flight. What had begun as a way to satisfy a fundamental aspect of the human imagination for millennia quickly developed into a commonplace technology that connected the world and made it smaller than ever thought possible. An unintended consequence of this shrinking is that dangers can also spread across the world in a matter of days, as was evident during the recent COVID-19 pandemic. The ability to access exotic materials around the world, allowed by a supply chain supported by aircraft, and the drive to further improve existing technologies has brought us to the current frontier of flight technology: hypersonic vehicles, capable of traversing a continent in minutes. As we master the unique problems posed by this new technology, we will reduce travel times around the world to the limits of what is possible and open new avenues of access to space. But this will also create new ways of threatening security and undermining defense systems around the globe.

The progression from the steam engine to the bleeding edge of flight technology and the unintended consequences of each step in that sequence may cause one to question whether the benefits of technological progress are worth the new problems it inevitably causes. The counterweight to the proliferation of unforeseen problems caused by new discoveries is that while natural intuition and fortuitous circumstances allow us to develop intermittent, serendipitous breakthroughs that are seemingly unrelated to the existing state of the art, problems that have already been identified can be systematically solved using the rigorous application of principles taught during an engineer's education and training.

A race between two efforts often begins the moment we invent a new technology or find a new way of transforming energy. The first effort is to implement that new innovation in as widespread and effective a way as possible and to find the broadest possible applications for said innovation. The second effort is to develop an understanding of the underlying physical principles that make the new discovery possible. This is done in order to both improve the newfound technique or tool and find other ways that this new physical principle can be applied to different problems. Both the act of applying a new technology to far-ranging areas and the study of the underlying principles of a new technology allow for the possibility to generate more technologies, which then begins a new race. Thus, for each new technology or technique of energy management developed, many more may spawn after a period of time.

This competing effort that allows for the growth of new technologies can also be examined in the context of the philosophy of Heraclitus. In addition to his theory of fire being the core element of the world, the other major extant portion of his philosophy is his belief in the unity of opposites. To Heraclitus, a beginning was also an end, and major events were merely a point in a cycle that floated between two extremes, each feeding off of the other in perpetuity. This cycle of codependent yet opposite forces is manifest in the ways that we push the boundaries of what is possible. One of the basic human tendencies is to create and transform. Out of nothing, we construct tools, systems, machines, and works of art, often manifest in different objects and occasionally in a single form. There is no need to understand hydrodynamics to float a raft or statics to build a table. Our natural intuition is enough to guide us to practical solutions. An engineer's experience may make the path more clear, but the drive to create something new was inherent in humanity long before engineering was set apart as a field of its own. An oft-repeated definition of engineering is the process of solving a problem with the use of scientific and mathematic principles. Yet oftentimes we create things that are so groundbreaking that there does not yet exist a theoretical framework sufficient to describe the function of the creation; in such cases our drive to create has led us to take a step beyond the precipice of existing knowledge. According to the previously stated definition, this, then, is not engineering. Once we've taken a step out into the void, we then find ourselves quickly constructing a platform under our feet so that we need not step backward into the previous status quo. This is where engineering begins to play its part and is the essence of engineering research: to understand the underlying principles of devices we've produced out of natural intuition and fortunate circumstances in order to intelligently improve upon the initial design. Spontaneous creation and deliberate analysis work in tandem to push our understanding to new boundaries while simultaneously building scaffolds of understanding that allow us to branch off into new directions of inquiry.

The development of new disciplines after a new invention is a process repeated frequently since the industrial revolution. The steam engine ushered in thermodynamics, the airplane created aerodynamics, and the computer changed linear algebra from a subdiscipline of geometry to one of the fundamental subjects in modern math education. In each of these examples a new technology or way of transforming energy was discovered, and a new field of research was created in order to either better understand the underlying physical phenomena that allowed that advancement to occur or better utilize the technology that had been created. The extraction of energy through pistons and turbines does not exist in nature and never would have occurred if not by human enterprise; therefore, the study of energy transfer was inconsequential until a device that utilized that process was created. The creation of steam engines required the development of a way to analyze how energy was moved and transformed throughout a system in order to make these machines more efficient and useful, and thus the discipline of thermodynamics was created. The impact of new analysis techniques extends beyond the improvement of the original device of interest, as these new techniques also provide a new lens through which the natural world can be viewed. The fundamentals of thermodynamics may also be applied to other fields, such as biology. This merging of fields allows life to be viewed as not only an act of survival but also a competition for energy. As technological progress continues to accelerate over the next 50 years and new analytical frameworks are required to facilitate new developments, we will also gain new ways of understanding the world around us.

To predict what the state of engineering and the world will be in 50 years would be an act of the imagination, influenced by perceptions about problems facing the world and the limits of certain fields of science and engineering. A science fiction author could likely make a guess as accurate, if not more so, as a prediction made by an eminent technical expert. What is certain is that spontaneous breakthroughs ushered in by strokes of genius and fortune will lead us into exciting, uncertain, and sometimes dangerous new territories. Meanwhile, the persistent and rigorous application of engineering fundamentals will allow us to understand the new states we find ourselves in, alleviate unforeseen risks, and build a jumping-off point for the next great advancement.

12

ENGINEERING 50 YEARS FROM TODAY

EVA

Eva is a fifth grader in Mrs. Lehman's class.

Engineering 50 years from today

Lehman 4-5   Eva Richards #21

I think in 50 years, engineering
for spaceships will be improved alot.
By that time I think if they are
improved, people will be able to
travel into other solar systems or
further into space. they will see
new planets and maybe even discover
plant life or other life in 50 years.

I also think that there will
be more robotics and AI. Robots
and AI may make things alot easier
to do stuff but much of it we
can just do ourselves.

Lastly I think in 50 years
cars and other things that take you
places will be alot easier to
control and use. People right now
are working on how to make self
driving cars too.

13

MANUFACTURING THE FUTURE: A 50-YEAR OUTLOOK ON AERO/ASTRO ENGINEERING FROM A PURDUE GRAD

DANIEL GILLIES

Daniel Gillies earned his degree in aero/astro engineering from Purdue University in 2006. His engineering career includes work on the Space Shuttle and the International Space Station, SpaceX's Falcon and Dragon, NASA's Stratospheric Observatory for Infrared Astronomy, the National Oceanic and Atmospheric Administration's geostationary weather satellites GOES-R and GeoXO, Boeing's 787 Dreamliner and Chinook, and Astrobotic's Griffin Lunar Lander.

I n my line of work, supporting business development and engineering for aerospace startups, I constantly am thinking about what the future holds for our discipline. Recently this has become increasingly difficult, with the industry as a whole having hit its "hockey stick" moment, a point in growth where gains move from being linear to exponential, resulting in the validity of projections of "what's next" expiring faster than ever before. In retrospect, when I was taking classes at Purdue, this was not the case. At the start of my Purdue tenure in the summer of 2003, it even seemed that a golden era of aerospace development and progress was behind us. Revolutionary vehicles such as the Concorde were taking to the sky for a final time, and the promise of Mars was always 20–30 years away. There was significant focus on optimization, perhaps more for increasing large aerospace developer and operator shareholder value and much less on revolutionary designs and rapid growth.

By the time I was getting ready to graduate in 2006, there was the promise of new growth, but it was still in its infancy. SpaceX was making its first attempt at launching Falcon 1. I remember a group of us huddled in the aero/astro engineering computer lab in Grissom Hall watching the first attempt. Little did we know where that flight would lead today. In parallel we were offered the promise of the National Aeronautics and Space Administration (NASA) Constellation Program to return boots to the moon. Many of us dreamed of our chance to work on programs such as the Altair Lunar Lander, perhaps giving us our own Thomas Kelly moment, a key engineer and program manager on the Apollo Lunar Module. This promise was short-lived, with not only Constellation eventually drying on the vine but also revolutionary programs such as the NASA Institute for Advanced Concepts shutting their doors in 2007, temporarily closing a chapter on NASA's own push for the next great new idea.

However, today as a discipline, aeronautical and astronautical engineering is a rapidly accelerating wild frontier. In aeronautics, we have seen developments ranging from all-composite computer-aided designed aircraft and the rebirth of commercial supersonic travel to hypersonic flight test programs becoming almost commonplace. In astronautics, reusable commercial rockets (Falcon 9, Falcon Heavy) and even spacecraft (Dragon) have become the expectation, not the exception, and commercial companies such as Intuitive Machines, Firefly, iSpace, and Astrobotic have shown that it is not only governments that can push outward to the moon and beyond. We are now rapidly beginning to ascend on our hockey stick trajectory, and what comes ahead of us has become increasing challenging to predict.

As we look forward to the next 50 years of growth in aeronautical and astronautical engineering, the growth of industry in space will be the driver of the next wave of innovations and endeavors. While the space economy today has revolved around access to and operations within Earth's orbit, the expansion of applications in LEO to GEO will likely plateau. Resources in this space are already limited, marked by crowded orbits with ever-increasing risks of collision, with the applications themselves such as communications and Earth observation experiencing further congested marketplaces. Instead, true growth in space is likely to come beyond Earth's orbit and wherever new resources are available, be that on the moon, from asteroids, or even from Mars. In the next two decades, I expect that resource exploitation from these destinations will begin to take hold, with exponential consumption driven by in-space manufacturing and utilization. Just as reusable rockets are the norm today, I expect that in 50 years we will see in-space resource collection and subsequent manufacturing become equally as normal. From this new industries will bloom, and as a result the hockey stick growth we are seeing today will not plateau; instead, it will continue to soar as space resource utilization takes off in every possible direction, all away from Earth.

To support this industrious future, I expect near-term investments in technologies that will enable access to these distant resources. Larger launch vehicle capabilities will give birth to spacecraft capable of carrying greater quantities of propellant—providing the boost to access resources beyond GEO and even the moon—within the capabilities of commercial companies. When you consider the vast investments made terrestrially in resource exploration, such as establishing a new oil well, I suspect that we will see equal if not greater investments in space resource exploration when it is proven that the technology to access the resources has become commercially available. Initially, the ability to return materials from these locations will also be a significant technological driver. However, within the next 50 years this will be outpaced by in situ utilization and production. Significant growth in in-space manufacturing technologies will need to occur beyond the simple 3D printed accomplishments being made on-orbit today. Raw material processing and refining will be necessary, along with significant developments in power generation, solar or nuclear, to power these industrial processes.

Coming back to Earth, I also do not expect aeronautics to slow, given the growth seen presently in both speed and mobility solutions. I expect supersonic commercial transport to become the norm. Initially it will fill the same void left by Concorde, catering to a more exclusive clientele but over time trickling down to serve the masses. More personalized air transportation will become widely available through the use of automation, with air taxis becoming as commonplace as an Uber ride today. Both of these advancements will be made possible through more efficient airframes and power plants, especially advancements in energy storage technologies.

These outlooks are made from the perspective of the present and are unavoidably biased by current trends and technology directions in aeronautics and astronautics. However, in space I fully believe that growth will be driven by a demand for further resources, while on Earth I see aeronautics pushing forward and being driven by increased needs for on-demand mobility in an already congested world. Future Purdue engineers, with even greater vision, will determine the true reality of these outlooks through their cutting-edge research being performed on campus and in the industrious companies that Purdue grads will create and grow to turn ideas into products and new industries.

14

ENGINEERING IN 50 YEARS: THE EVOLUTION OF INNOVATION AND EDUCATION

YAERID JACOB

Yaerid Jacob is the founder and CEO of Blueprint Data Centers. Over the course of his career, he has executed multiple $100 million to $1 billion-plus energy and real estate projects as a developer, consultant (design and project management), contractor, and investment banker.

The thought of engineering in 50 years is both exciting and humbling. As I reflected on this topic, I reached out to several of my former classmates—leaders and innovators in various engineering disciplines—to gather their perspectives. The consensus was clear: engineering will continue to be a driving force in shaping the world, but the way we approach it, the skills required, and the role of education in preparing future engineers must evolve significantly.

THE FUTURE OF ENGINEERING AND PURDUE'S ROLE

For 150 years, Purdue University has been a beacon for those eager to build, innovate, and solve real-world problems. This legacy will persist as long as Purdue remains a place where individuals who are passionate about engineering can come to learn and create. The university's global impact will be measured by its ability to adapt, ensuring that future engineers are equipped with the right skills to lead in an era defined by rapid technological advancements.

The world of engineering in 50 years will be vastly different from today. Automation, artificial intelligence (AI), and potentially quantum computing are just some of the fields that will redefine engineering. The need for interdisciplinary knowledge will grow as engineers will be expected to collaborate with professionals from other professions and engineering disciplines to address global challenges. Purdue must not only keep pace with these changes but also pioneer new educational methodologies that prepare students for careers in fields that may not even exist today.

BRIDGING TODAY WITH THE NEXT 50 YEARS

The next five decades will bring profound changes in how engineers operate. The days of mastering every technical detail are fading. Instead, the engineers of tomorrow will excel by leveraging advanced tools, software, and automation. Employers will increasingly integrate AI-driven design, simulation, and project management tools to streamline workflows, thereby reducing manual tasks and reallocating human talent to quality assurance, project leadership, and innovation.

This shift necessitates a change in engineering education. Future engineers must be fluent in the language of evolving technologies in addition to cultivating strong communication skills and adaptability. Companies will look for engineers who not only understand technical principles but can also lead cross-functional teams and integrate rapidly evolving technology solutions.

Purdue has historically prepared its students well for the workplace by blending theoretical knowledge with practical experience. I and many of my peers found that this approach gave us a great starting point in our first full-time jobs over some of the graduates from other institutions that were heavily technical but provided limited practical experience. As the industry evolves, this balance will be even more crucial in shaping competent engineers.

ENGINEERING EDUCATION: BALANCING TECHNICAL AND PRACTICAL SKILLS

One of the most significant challenges for future engineers will be bridging the gap between technical expertise and practical application. Being technically strong is invaluable, but the true test lies in execution: how effectively engineers can translate their knowledge into impactful solutions.

Courses such as those taught by Dr. Neal Houze at Purdue were instrumental in highlighting this balance. These classes didn't just teach engineering concepts; they also instilled problem-solving abilities, adaptability, and the confidence to execute ideas in real-world settings. Expanding such hands-on courses will be key to preparing students for an ever-changing technological landscape.

Additionally, the depth of knowledge required for problem-solving is shifting. Previous generations may have needed extensive foundational knowledge to solve a problem, whereas today advanced tools can significantly reduce the cognitive load required. While past engineers may have needed 10 different pieces of information to solve a challenge, future engineers might only need two or three as long as they know how to effectively use the tools at their disposal.

Furthermore, soft skills will become just as important as technical abilities. Engineers must develop skills in leadership, teamwork, and creative problem-solving. With the rise of remote work and globalized projects, the ability to communicate ideas across different cultures and time zones will be a crucial differentiator. Universities must recognize this shift and embed more opportunities for collaborative projects, internships, and real-world applications into the curriculum.

A DUAL APPROACH TO ENGINEERING EDUCATION

Students entering higher education often have little clarity about which path within engineering suits them best. Some thrive in deeply technical, research-driven environments, while others excel in hands-on, application-based roles. The future of engineering education could benefit from a dual-track approach.

In this model, the first two years would offer a broad engineering foundation for all students. Following this, students would have the choice to specialize in one of two streams:

- A highly technical and research-focused track for those interested in pushing the boundaries of engineering theory and innovation, or
- A practical, industry-oriented track designed to prepare students with the tools and methodologies necessary to thrive in the workforce.

Such an approach would allow students to discover their strengths and interests before committing to a specific trajectory and would also ensure that both industry and academia receive graduates with the right skills for their respective domains.

Moreover, lifelong learning will become a necessity rather than an option. As technology evolves rapidly, engineers must continuously update their knowledge through professional development, certifications and continued education programs. Universities such as Purdue could lead in this space by offering flexible, ongoing learning opportunities for alumni to stay ahead in their fields.

A MESSAGE TO FUTURE ENGINEERS

One of the most important lessons I've learned is that being at the top of the class is not the only measure of success in engineering. Many students who may not excel in pure theoretical coursework find their true strengths in practical application.

For aspiring engineers who may struggle with the traditional academic model, it is essential to recognize that engineering is about more than just grades (even though grades will need to remain important). Engineering is also about solving problems, creating solutions, and making an impact. If technical theory doesn't come naturally, this does not mean that engineering isn't for you; it simply means that your approach to engineering may be more hands-on and execution-focused.

The engineering world needs both visionaries and doers—those who develop theories and those who turn them into reality. Whether an engineer thrives in research, innovation, or implementation, there is a place for everyone in this ever-expanding field. The key is to remain curious, adaptable, and committed to lifelong learning.

The future of engineering is bright, and Purdue University will continue to play a pivotal role in shaping the next generation of innovators. By embracing change, integrating new learning models, and recognizing the diverse strengths of students, we can ensure that engineering remains a driving force in shaping the world for the next 50 years and beyond.

As we look to the future, we must remain focused on the core purpose of engineering: to improve lives, drive progress, and create a better world. The next generation of engineers—shaped by institutions such as Purdue—will be the architects of tomorrow's breakthroughs, and it is our responsibility to equip them with the knowledge, tools, and mindset needed to thrive.

15

ENGINEERING 50 YEARS FROM TODAY

PIPER

Piper is a fifth grader in Mrs. Lehman's class.

Piper 20
Mrs. Lehman

Engineering 50 years from today

I think in 50 years medicine will change a lot over the years. I believe we will have a lot more medicen. I think we will have the cure to Cancer and most of the illneses the are a big problem. Their will hopefully be more people researching medicen just in case if a new desiese would acure.

I hope kidnaping will not be a thing, but it probably will. I was thinking what if you just put a chip in your babies like you chip your dog. Or what if you haft to text a number and if you don't text than they will look for you because they think you will have ben missing.

Technology will probubly be very futuristic and will probubly have flying cars. In the future we will probubly have I phone 35 ultra pro max. And we will probubly have flying cars and so much more.

16

EDUCATING THE ENGINEER OF 2075

LEAH H. JAMIESON

Leah H. Jamieson is the Ransburg Distinguished Professor of Electrical and Computer Engineering at Purdue University, the John Edwardson Dean Emerita of Engineering, and a Professor by Courtesy in Purdue's School of Engineering Education.

INTRODUCTION: PERSPECTIVES ON CHANGE IN ENGINEERING EDUCATION

Change in engineering education as a change agent, catalyst, advocate, and champion has been a part of my portfolio and persona for the past 30+ years. To frame change over the next 50 years, I profile three past transformations as a means of exploring potential drivers of change. I also look at the pace of change of engineering knowledge and what this might mean as we try to predict the future. I identify key attributes of engineering education in 2025 as a basis for postulating future attributes. And then I try to do the impossible: predict what engineering education will look like in 2075.

WHAT DOES SEISMIC CHANGE IN ENGINEERING EDUCATION LOOK LIKE? THREE EXAMPLES

TRANSFORMATION #1: FROM PRACTICE TO THEORY

With the establishment of land-grant colleges and universities under the Morrill Act of 1862, engineering education in the United States shifted from apprenticeships to classrooms. This ushered in debates about the balance of practice and theory. The definitive shift to theory occurred in the aftermath of World War II, when the contributions of engineers were seen as far less significant than the contributions of physicists. Called "the most significant change in engineering education during the past 100 years" (Froyd et al., 2012), the movement to engineering science as the defining core of educating engineers was aided by the creation of the National Science Foundation to fund science research, also an outgrowth of World War II (Bush, 1945). The embracing of engineering science and theory as the cores of engineering education was codified by the American Society for Engineering Education Grinter Report (Grinter, 1955) and dominated engineering education into the closing decade of the 20th century.

KEY DRIVER OF CHANGE:

- Perceptions of shortcomings of engineers in World War II.

TRANSFORMATION #2: FROM ENGINEERING SCIENCE TO LEARNING BY DOING

The shift away from engineering science as the foundation of engineering education had its roots in the early 1990s. Industry was slamming both the Accreditation Board for Engineering and Technology (ABET) and engineering degree programs for weak accountability and attention to quality and for failing to keep up with the industry's changing expectations for newly hired engineers. At Purdue, for example, this call to action directly inspired creation of the Engineering Projects in Community Service program in 1995 (Coyle et al., 2006).

Under these pressures, the pendulum that had swung so decisively toward engineering science began to reverse its direction, restoring the art and craft of engineering design to the curriculum, along with the professional skills that support high-quality design. ABET's Engineering Criteria 2000 brought outcomes-based accreditation that included communication and teamwork, ethics in a global, social, intellectual, and technological context, lifelong learning, and synthesis of engineering, business, and societal perspectives. The National Academy of Engineering (NAE) reinforced this sea change with its "attributes of engineers in of 2020," including practical ingenuity, creativity, professionalism, and dynamism, agility, resilience, and flexibility (National Academy of Engineering, 2004). The pendulum swing to engineering practice significantly broadened the ways to experience engineering beyond traditional industry co-ops and internships, adding undergraduate research and entrepreneurship experiences, community-based projects, and global engagement as potential components of engineering students' education.

We are about 35 years into the experiential education era and have learned that experiencing engineering (including computing) is doing a better job of meeting industry's needs. An unanticipated side effect has been that engineering experiences early in a student's academic career also have a positive impact on the retention of students, especially women, in their early years in engineering.

KEY DRIVERS OF CHANGE:

- Criticism from industry that engineering education was not keeping up with evolving industry needs.
- Criticism from industry and academic leaders that accreditation was not assessing quality.
- NAE's call to prepare for the future of engineering rather than react to change as or after it happens.

TRANSFORMATION #3: COMPUTING AND COMMUNICATION TECHNOLOGIES

Since the 1960s, computing and communication have continually changed the face of engineering education. The first Department of Computer Science in the United States was launched at Purdue University in 1962. Computer engineering emerged as a separate discipline in the 1970s, with the first computer and electrical engineering degree established in 1980. The impact of advances in these technologies went far beyond the new computer science and computer engineering disciplines: computing and communication tools have transformed virtually all disciplines. With exponential increases in computing power, simulation has joined theory and experimentation as a third pillar of problem-solving. By enabling virtual teaming, the internet has transformed collaboration and cooperative learning and has fundamentally changed the delivery mechanisms for learning, ranging from web searches to online courses and virtual/augmented reality tools. At the same time, the ubiquitous low-cost internet has enabled massive repositories of examples, problems, and solutions that have disrupted education by challenging educators to rethink assessment of learning. Computing and digital communication are a part of education in all engineering disciplines (and many or most disciplines outside engineering) and are essential components of job readiness for engineering students. But the computing/communication story doesn't end here. The most recent chapters in speech understanding, large language models, deep learning, data science, and artificial intelligence are springboards to the future.

KEY DRIVERS OF CHANGE:

- Key leaders' recognition of the potential impact of computing leading to the creation of computer science and computer engineering departments and degrees.
- Pervasiveness of computing and the internet in education and industry.
- Continual advances in computing and communications technologies.

TIMESCALES FOR CHANGE

What does 50 years look like? The engineering science era lasted on the order of 50 years, from approximately 1945 to 1995. The engineering design/ experiential education era is roughly in its mid-30s and appears to have staying power for at least a while longer. The steady changes driven by computing and communications technology began in the 1960s and have spanned at least five decades. Early in the 2020s, generative artificial intelligence (AI) exploded onto the scene in unprecedented ways, signaling the start of a new seismic change in many dimensions, including engineering education. Generative AI has also challenged our thinking about the pace of change in engineering.

One measure of the pace of change of engineering knowledge is the half-life: the time span over which half of an engineer's knowledge will become obsolete (Charette, 2013). Estimates put the half-life of a vintage 1930s engineering degree at 35 years. This had dropped to 10 years by the 1960s, 5 years by 2007, and 3 years by 2023. The half-life of machine learning knowledge is currently estimated at 2 years and falling (McKendrick, 2024). This offers a good (but intimidating) jumping-off point for turning our attention from the present to the future.

ENGINEERING AND ENGINEERING EDUCATION IN 2075

It's 2075. What does engineering education look like? For that matter, are there still engineers? Are they humans? Computers? Robots? What do they do? And what does their education need to look like?

Change in engineering education has continued to be driven by both advances in technology and changing perceptions of the shortcomings of engineers. All fields of engineering have evolved significantly, but the seismic change that erupted in the form of large language models and generative AI in the early 2020s has played a leading role across all of engineering. Following the initial onslaught of activity spurred by ChatGPT (Manjoo, 2022; Roose, 2022), change has come in bursts, but the estimated two-year half-life has stabilized rather than shrank to fractions of seconds. Like the 50-year progression in computing from the 1960s to the 2010s, the change over the period 2025–2075 period has been both gradual and dramatic, and while technology has been a consistent driver of change, human perceptions and needs have also shaped what engineering education looks like in 2075.

Aside: The path to 2075 could play out in a number of ways. For example,

- **Dystopian alternative future:** AI, autonomy, and the Internet of Things have combined to destroy society, leading to a tech backlash, so 2075 resembles the pre–Industrial Revolution era.
- **Who needs education?** All engineering is done by computers/sensors/robots, so there is no need to educate engineers. Humans can sit back and enjoy their lotus-eater leisure—except for those humans who thrive on learning, doing, conquering challenges, having a purpose, and making an impact.
- **Exquisite symbiosis:** Through research, inspiration, and necessity, humans and AIs tune their partnerships to take advantage of their respective strengths, including learning how to value the contributions of each (Castelvecchi, 2024).

I prefer to be an optimist, so the future I explore is engineering education in the world of high-performing human-AI symbiosis.

To envision engineering education in 2075, we first need to envision engineering. Memory, speed, and sensory input/output capabilities are beyond current imagining, as are seemingly limitless communication capabilities. The AI of 2075 will have access to all past data and engineering knowledge and the ability to synthesize across all its knowledge.

With the technical base and encyclopedia of all past engineering experience at the AI's virtual fingertips, the role of the human engineer has changed.

In the era of the omniscient AI, the engineer's bread-and-butter role is one of assessment: triaging problems that need engineering solutions, assessing relative merits of potential solutions generated by AIs, applying human lenses to evaluating the performance and effectiveness of fielded or experimental engineered systems, and, broadly, combining the roles of design reviewer and Socratic questioner of the AI engineer.

Beyond these new routine roles, engineers continue to be, as they have always been, creative imaginers and discoverers of ideas never before considered. Humans and machines together make new discoveries and develop new technologies and new environments in areas where the knowledge base is still in progress, such as human habitats undersea and on Mars, new science emerging from the measurement of new galactic phenomena, and new settings for learning and leisure. The engineer is an explorer of the frontier created by needs and opportunities that have risen from contexts so new they are not yet in the AI's engineering knowledge omnibus. In partnership with thinkers and doers from the arts, sciences, and social sciences, the engineer is a seeker-outer, an interpreter, a champion, and a translator of yet-to-be-explored and uniquely human needs and dreams. Engineers bring their unique creativity to vibrant partnerships with AIs to tackle these new challenges.

These changes in what engineers do have driven changes in engineering education. Early in the 50-year window, lifelong learning has become the norm. Over the decades leading up to 2075, AI-enabled personalized education has succeeded in shifting the focus of education from teaching to learning and has prepared engineering students for increasingly substantive interactions with AIs. Learning delivery and interaction technologies include well-established natural language, virtual reality, and augmented reality interactions as well as the more recent brain-to-computer and brain-to-brain communication. The primary emphasis in engineering education is not on what you know but on how you think. And while the broad purpose of engineering education has remained fundamentally the same—to educate the next generations of both engineers and engineering educators who will be involved in "advancing technology for the benefit of humanity" (IEEE, 2020)—the details have changed significantly. Both engineering and engineering education are now a deep partnership—the exquisite symbiosis—between humans and AIs and between technology and society.

How engineering education has changed becomes concrete when we compare key criteria for the engineer of 2075 with some of the ABET

2024 criteria and curriculum requirements and the NAE's "attributes of engineer in 2020."

- The "ability to . . . solve complex engineering problems by applying principles of engineering, science, and mathematics" (ABET, 2024) has been superseded by much-expanded emphases on communication, teamwork, empathy, and social science, with the ability to partner with both human and machine collaborators now being a foundation of engineering education.
- With increased dependence on AI and public mistrust around issues that were prominent in the 2020s—image, video, and audio manipulation and fabrication; reasoning beyond the data; and transparency, bias, and decision-making—ethics as it specifically relates to machine learning, autonomy, decision-making, and privacy now plays a prominent role in engineering education and accreditation. While the issues of the 2020s have been resolved through a combination of AI research, policies, and regulations, ABET 2024's "ability to recognize ethical and professional responsibilities in engineering situations" now explicitly includes partnering with technology and calls out the responsibility of the institution in connecting engineering with ethics and the social sciences. Pressure is increasing to consider ethical issues in parallel with the development of new technologies rather than after the fact, including debates about whether AIs and robots are now part of society.
- The ABET 2075 criteria include "alignment," which AI experts describe as "making sure A.I. systems are in line with human values and goals" (Metz, 2023).
- ABET's "culminating major engineering design experience" is still a defining achievement of engineering students, but it now explicitly measures students' ability to take full advantage of AI's engineering knowledge base, their ability to collaborate appropriately with the AI as evaluated by both the AIs and humans advising their capstone experience, and their ability to demonstrate the still-human attributes of curiosity and creativity in defining and executing the project.

Most radically, human students and AIs are genuinely partners throughout engineering education, accreditation, and practice.

CONCLUSION

Engineering education's 50-year journey from 2025 to 2075 has, like the 50-year journey that preceded it, had computing and digital communication technologies as epoch-long drivers of change. Human assessments of how engineering and computing education are performing in their broader contexts have also been a part of the change, with a much-increased focus on ethics and empathy and on both the quantitative and qualitative effectiveness of the human-AI partnership.

Looking both back and forward, over the 50-year windows much has changed. However, two things have not changed. Engineering is an integral part of our lives. And in the spirit of optimism, in 2075, as in 2025, it's a great time to be an engineer, especially a Purdue engineer.

REFERENCES

ABET. (2024). *Criteria for accrediting engineering programs, 2024–2025.* https://www.abet.org/wp-content/uploads/2023/05/2024-2025_EAC_Criteria.pdf

Bush, V. (1945). *Science, the endless frontier: A report to the president on a program for postwar scientific research.* Office of Scientific Research and Development.

Castelvecchi, D. (2024). Huge randomized trial of AI boosts discovery—At least for good scientists. *Nature, 636*(8042), 286–287. https://doi.org/10.1038/d41586-024-03939-5

Charette, R. N. (2013, September 4). *An engineering career: Only a young person's game?* IEEE Spectrum. https://spectrum.ieee.org/an-engineering-career-only-a-young-persons-game

Coyle, E. J., Jamieson, L. H., & Oakes, W. C. (2006). 2005 Bernard M. Gordon Prize Lecture: Integrating engineering education and community service: Themes for the Future of Engineering Education. *Journal of Engineering Education, 95*(1), 7–11. https://doi.org/10.1002/j.2168-9830.2006.tb00873.x

Froyd, J. E., Wankat, P. C., & Smith, K. A. (2012). Five major shifts in 100 Years of engineering education. *Proceedings of the IEEE, 100* (Special Centennial Issue), 1344–1360. https://doi.org/10.1109/JPROC.2012.2190167

Grinter, L. E. (1955). Report on evaluation of engineering education. *Journal of Engineering Education, 46*(1), 25–60.

IEEE. (2020). *IEEE Strategic Plan 2020–2025.* https://www.ieee.org/about/ieee-strategic-plan.html

Manjoo, F. (2022, December 16). Opinion: ChatGPT has a devastating sense of humor. *New York Times.* https://www.nytimes.com/2022/12/16/opinion/conversation-with-chatgpt.html

McKendrick, J. (2024, April 30). AI puts the squeeze on the shrinking half-life of skills. *Forbes.* https://www.forbes.com/sites/joemckendrick/2024/04/30/ai-puts-the-squeeze-on-the-shrinking-half-life-of-skills/

Metz, C. (2023, March 31). What's the future for A.I.? *New York Times.* https://www.nytimes.com/2023/03/31/technology/ai-chatbots-benefits-dangers.html

National Academy of Engineering. (2004). *The engineer of 2020: Visions of engineering in the new century.* National Academies Press. https://nap.nationalacademies.org/catalog/10999/the-engineer-of-2020-visions-of-engineering-in-the-new

Roose, K. (2022, December 5). The brilliance and weirdness of ChatGPT. *New York Times.* https://www.nytimes.com/2022/12/05/technology/chatgpt-ai-twitter.html

17

T², THE FORMULA FOR ENGINEERING THE FUTURE

A Boilermaker's Journey from Shop Floor to Global Stage

KEITH KRACH

Purdue engineer **Keith Krach** founded several transformational companies and served as Under Secretary of State. His mission has always been to lead with integrity, build with trust, and empower others to take bold leaps. As Krach puts it, "Transformation to the power of Trust—T²—is the magic formula for building a better world."

THE SPARK: FINDING COURAGE AT PURDUE

Fifty years ago, I walked the halls of Purdue University as a wide-eyed industrial engineering student, not fully grasping how profoundly this place would shape my life. Back then I didn't know what "transformation" meant, and I certainly didn't know that one day I would describe it as an equation. But I felt it. Purdue didn't just teach me systems analysis or control theory—it lit a fire in me. Purdue gave me the courage to jump into water over my head and trust that I wouldn't drown.

Purdue taught me something foundational: that trust is the coin of the realm, and when you take transformation to the power of trust, or what I call T^2, that's when the magic happens.

I grew up welding parts in my dad's five-man machine shop in small-town Ohio. His dream was for me to get some "college knowledge," come back as an engineer, and help him grow the business into a big company of 10. I never returned to work in that shop, but he was proud that I became a Boilermaker—and especially proud that it was also the name of his favorite after-work adult beverage.

But what he never knew was how terrified I was when I arrived at Purdue. I wasn't in the honors classes like my classmates. I was intimidated, overwhelmed, and outgunned. I remember sitting in Engineering 100 and hearing Professor James W. Barany say "Look to your left, look to your right—one of you won't be here next year." I was sure he meant me. But something inside me shifted. That fear became a motivator. I learned to jump into the unknown, to transform myself—and to trust that I'd swim.

FROM WELDING TO WORLD STAGE: THE T² EQUATION IN ACTION

That instinct—to leap, to transform, to build trust—became the cornerstone of my life's work. Every company I've built or led was grounded in the belief that trust is the foundation of transformation.

At GMF Robotics, we helped General Motors become the largest industrial robotics manufacturer in the world—a transformational leap in American manufacturing. But we weren't just selling machines. We were convincing factory managers to trust machines not to hurt their workers, to trust new workflows that hadn't yet been proven, and to trust our small team to deliver something the world hadn't seen before.

At Rasna, we reinvented how engineers designed mechanical systems. We weren't just offering software—we were asking people to trust an entirely new way of thinking about physics, computation, and design.

At Ariba, we created the world's first business-to-business e-commerce network. When we started, digital procurement was a foreign concept. Twenty years later, the Ariba Network transacts more than $6 trillion annually—more than Amazon, eBay, and Alibaba combined. But that kind of growth only happens when companies trust you with their supply chains, their data, and their reputations.

And at DocuSign, we didn't just digitize signatures—we redefined trust in the digital age. I used to tell our team, "We're not in the software business. We're in the trust business." After all, people use DocuSign for their most important agreements—the ones they sign. Today, DocuSign serves over a billion users and a million companies, and the name itself has become a verb. Why? Because people trust it.

In every case, the equation held true: Transformation raised to the power of Trust = Impact.

TECH STATECRAFT: WHEN TRUST BECOMES POLICY

That same equation guided me when I entered public service. Leaving Silicon Valley to serve as Under Secretary of State for economic growth, energy, and the environment was a leap into deep waters but one rooted in purpose.

When I was tasked with crafting a global economic security strategy, it became clear that the battle wasn't just economic—it was also ideological. Authoritarian regimes were using technology to control populations, surveil citizens, and suppress freedom. And while many of our allies didn't want to say it publicly, when asked how their relationships were going with these regimes, they would whisper "We don't trust them."

That insight became the basis for a new kind of diplomacy—tech statecraft—that fused Silicon Valley speed and strategy with foreign policy values. We built a global alliance based not on coercion but instead on shared trust principles: integrity, transparency, reciprocity, and respect for human rights, the rule of law, property rights, national sovereignty, and the environment.

To carry that mission forward, I partnered with Purdue University president Mung Chiang, a brilliant mind I recruited into the State Department while he was the dean of engineering. Together, we launched the Krach Institute for Tech Diplomacy at Purdue, which has become the world's preeminent institution advancing freedom through trusted technology. This is a powerful reminder of what happens when academia, business, and diplomacy come together with a shared purpose.

ENGINEERING THE FUTURE WITH PURPOSE

Looking ahead, the future of engineering will be shaped not just by how fast we innovate but also by the values we embed in that innovation. Technology is no longer just about performance—it is about purpose. Technology is about who builds it, how it is built, and what it is built for.

We must engineer for not only efficiency or elegance but also resilience, accountability, and dignity. That is what trusted technology means. It is not enough to build things that work—we must build things that people believe in: systems that are secure, algorithms that are fair, and infrastructures that are open, ethical, and free from coercion.

Artificial intelligence (AI) will revolutionize not only what we build but also how we build it. AI is fast becoming more than a tool—it is a collaborator. Engineers of the future won't just write code or solve equations; they'll orchestrate ecosystems, partnering with intelligent systems to imagine, iterate, and execute in real time.

But again, it comes back to trust. Who programs these systems? What data are they trained on? Are the outputs transparent and explainable or opaque and biased? As engineers, it will be our responsibility to ensure that AI is developed ethically, inclusively, and in service of humanity. Just because AI can do something doesn't mean that it should.

The engineers of the next 50 years must be as fluent in ethics as they are in equations. That's what it means to lead in the era of T^2.

THE PURDUE EQUATION: A LEGACY TO LIVE

As I look back at that young student who first walked across Purdue's campus, uncertain but determined, I realize now that Purdue gave me more than an education—it gave me a lifelong equation:

$$T^2 = \text{Transformation to the power of Trust}$$

This is an equation that helped me build companies, lead teams, navigate government, and even shift global diplomacy. It is an equation that belongs not to me but to every Purdue engineer who dares to make a meaningful difference.

The next 50 years will belong to those who are bold enough to jump in water over their head and wise enough to bring others with them.

To the engineers of the future, trust yourselves, transform fear into fuel, and never stop building.

And so, to the future engineers who will shape this world, remember that trust is your blueprint, and transformation is your tool kit. Use them boldly. Use them wisely. And never forget: the most powerful equation you'll ever engineer isn't written in any textbook. It is the one you live.

T^2 isn't just a formula. It is a philosophy, and it starts right here at Purdue University.

18

ENGINEERING
50 YEARS FROM TODAY

AUTUMN

Autumn is a fifth grader in Mrs. Lehman's class.

Technology already works so
fast today, imagion how fast
it will work in 50 years,
I guess we will wait and
see!

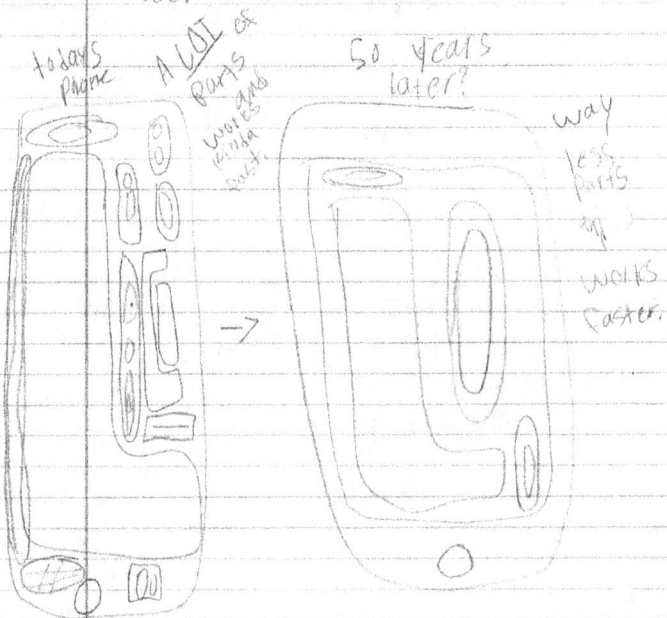

todays phone

A LOT of parts works kinda fast

50 years later?

way less parts

works faster.

19

PRESERVING PUBLIC TRUST IN TECHNOLOGY

RON M. LATANISION

Ron M. Latanision is the Neil Armstrong Distinguished Visiting Professor in the College of Engineering at Purdue University. He taught at MIT prior to joining the consulting firm Exponent as a corporate vice president and is currently the firm's first senior fellow.

I think more and more about how my vision and trust in technology has evolved and, importantly, how the public trust in technology seems now in the balance. All of those who are technologists must consider how to preserve that trust on the part of the public where technology serves social interests and needs.

Earlier this year and in the context of my role as the Neil Armstrong Distinguished Visiting Professor at Purdue University, I spoke at the university about a topic of intense interest to me, the Materials Genome Initiative (MGI), a product of the Office of Science and Technology Policy during the Obama administration. The MGI was launched in June 2011 with the goal of deploying advanced materials that meet national priorities and societal needs. Instead of trial and error, materials are being developed using first principles a part of this computational materials science and engineering enterprise with the aim of reducing the materials development cycle of 10–20 years by more than 50% and reducing the development cost proportionately. The MGI is reshaping materials science and engineering in terms of education and practice and is a demonstrably useful process by which technology serves society in terms of such materials as photoelectrodes for splitting water to produce hydrogen, a source of energy. Sunlight and water are the most abundant resources on our planet, and they are both free. They are available to legacy and nonlegacy nations. There are no political or geopolitical boundaries of concern.

I began by first telling my audience how honored I was to speak as a Neil Armstrong Professor. I never met Neil, but like millions of people around the world, I watched him step off the Apollo 11 lunar module onto the surface of the moon just before 11 p.m. EST on July 20, 1969. I was a bit younger then. Most of my audience had not been born yet! In 1969, I watched Neil and Buzz Aldrin together from our living room in Silver Spring, Maryland. I was a postdoc at the National Bureau of Standards in Gaithersburg. This is one of those moments that you never forget. I often think about the demonstration of trust in science and technology that Neil and his astronaut colleagues exhibited during this historic event and many others. And I felt a genuine sense of pride as an engineer in all of this. Just remarkable!

Although I had not met Neil, I did in fact work with astronauts while I was at MIT as the director of the Materials Processing Center. At that time, the National Aeronautics and Space Administration had a great deal of interest in microgravity processing of materials: large single-crystal semiconductors, protein crystals, and many others. My colleagues and I at the Materials Processing Center would visit the Marshall Space Flight Center in Huntsville periodically and help prepare and work with SkyLab and later SpaceLab astronauts. Space-grown crystals will probably never be commercially viable, but they are consistently of high quality. Once we understand that fundamental materials science, we can search for a way to reliably reproduce these crystals on Earth, in the presence of gravity.

And I did meet and know one of the last men to walk on the moon, Harrison (Jack) Schmitt. He and Gene Cernan landed the Apollo 17 lunar module on the moon in December 1972. I met Jack while I was on sabbatical from MIT in 1982–1983 as an adviser to the US House of Representatives Committee on Science and Technology and he was a US senator from New Mexico. He is a geologist and was the lunar module pilot. And while he got off the lunar module first in 1972, he also got back on first and became the next-to-last man to walk on the moon!

But at this point in history today and having watched the ease with which disinformation and misinformation can be produced and circulated using the internet and the World Wide Web and even more so with the explosive evolution of generative artificial intelligence (AI), I have great concern. A world of alternative reality has been produced using contemporary technology: fake news, fake videos, and more. Public distrust of politicians has been a long-standing concern, but the growing public distrust of science and technology is an even greater concern to me, given the technologically intense planet on which we live. In both cases, the distrust has been earned and is the root of many of our social problems today. For example, we technologists have made it possible for fake news to become a regular part of our lives. Do we now need to complement this with generative AI–derived deepfake videos? Some technologists seem determined to induce total distrust of technology on the part of the public. But some thoughtful leaders, including Senator Chris Murphy (D-Conn) and Governor Spencer Cox (R-Utah), have launched a national conversation with the intent to restore the common good. They look to involving intellectuals and activists. Notably, technologists do not seem to be included in their thinking. I think that we should become part of the conversation.

We—technologists—developed the internet and then the World Wide Web. The World Wide Web was/is intended to provide a platform that would make information available globally, which is useful. I am certain that the developers had not intended that these otherwise useful technologies would be transformed into sources of misinformation and disinformation, among other distortions that have so affected our lives. This transformation is driven by fear and greed, not by interest in serving a useful social purpose, and (this is a long way of getting to my point) we technologists have not been a part of the conversation to assign responsibility and accountability to those who are misappropriating these technologies. The attitude seems to be that if it serves a political or geopolitical purpose, do it. Or, if it makes money, do it! I am not opposed to technologically creative people becoming wealthy. I would just prefer that this wealth be derived from societally beneficial technology. To quote my friend Janusz Tabis, "As we are only bound by the laws of nature, it is crucial that wisdom underlies our decision-making. After all, because we are intelligent and can do something it doesn't mean that we should."

I often think about the public reaction to the introduction of any new technology into the marketplace. Generative AI is not any new technology. This one is shattering. But I suppose that to the average thoughtful person, the telephone must have been shattering, just as the Model T was. What is different is the case of AI is that it does not just add a new dimension to our lives; it presents technology as a force beyond nature. AI allegedly thinks and feels, although it is not clear on what scale and in what detail it compares to human thinking, as we don't really understand the particulars of how humans think. I just worry a bit that technology may be heading so far out front of humans that people may begin to distrust science and technology on a level that is unprecedented today. In my opinion, that would be counterproductive beyond reason.

I think that in regard to medicine, communication, electric generation and storage, food production, air and water quality monitoring, public safety, and so on, above all AI does not diminish public trust in science and technology.

However, fear and greed have served as human motivators for a very long time Think about Sputnik. And think also about profit at the expense of all others attitude, the social media giants. Until we get beyond the greed stage—and I think of greed in terms of not only personal fortune but also personal agendas—technology will continue to appeal to some of society's most creative people for the wrong reasons. My overwhelming reaction is that the interface between technology and humans is vital in order to do this right.

Generative AI has value if used responsibly and with accountability. It has been useful in developing new materials from first principles, as in the case of the MGI. Computer modeling and simulation and data management are the key ingredients to this effort. Computational materials science and engineering is real. And such data management tools are valuable in other spaces as well. But I am equally confident that these tools can be and are being abused. I had hoped that we might have learned from the experience of the internet and the World Wide Web, but I am concerned that fear (in terms of losing market share) and greed (in terms of losing market share) will prevail in the tech giants and that this will allow abuse to overwhelm useful initiatives. Even more, this could irretrievably erode public trust in science and technology. I am really concerned on behalf of my children and grandchildren and generations to come that without responsible and accountable humans at the helm, generative AI may do just that. In sum, generative AI is potentially both supremely useful (medicine, protein research, new materials, etc.) and supremely dangerous. How it evolves into our social fabric is a function of the good or bad sense of those who are developing/presenting this rapidly evolving technology and those who choose to use it—for good or bad purposes.

Looking forward 50 years into their future and our collective future on this planet, it seems to me that in terms of both practice and education, corporate culture and engineering education in particular must change going forward. The only way to change corporate culture is to change the way the technologists we produce realize that they share responsibility and accountability for the technology they develop and introduce into the marketplace. We have struggled to introduce engineering ethics into the curriculum throughout the educational system. Likewise, the inclusion of social science into the engineering mindset in both practice and education is needed but almost absent.

It seems to me that the future of generative AI, for example, is all about how people will choose to use it: for good purposes or bad. I am confident that it will be used for both, just as is the case with the internet and the World Wide Web. But we seem to have learned nothing from the experience with such technology. The simple fact is that there are lots of people who will choose to use generative AI for bad purposes (e.g., scammers), and unless those who develop such technologies also develop a conscience and attach some limits (if that is possible) to what can be done with their technologies, we are in for a very uncomfortable ride.

In my judgment, we cannot expect this to occur through government policy changes, since there are not enough people in Congress who know about or are willing to listen to people who understand the impact of technology on people (for good and for bad). Nor can we expect corporations to self-police, which never works. But a change in corporate culture that considers equally the well-being of businesses' shareholders and customers would be a move in the right direction. I understand that for-profit businesses do need to profit in order to survive, and I am not opposed to corporate leaders and shareholders sharing in the profits, but I hope that at some point this will all come to benefit the public rather than businesses knowingly putting the public at risk.

20

THE FUTURE IS PHYSICAL

Engineering Intelligence into the World

YANG LEI

Dr. Yang Lei is a principal artificial intelligence/machine learning research engineer at HP Inc., where she builds real-world technologies that empower people, enhance experiences, and shape the future of intelligent living.

It was a hot summer night in 2005, just days after the national college entrance exam in China. My parents and I sat in the modest home of my high school teacher, Mr. Guoqiang Shao, anxiously discussing my future. "My advice is to learn a skill," he said with a serious tone that has stayed with me ever since. That simple but profound suggestion led me to choose electrical engineering as my college major, which is an inflection point that would shape my entire life.

Looking back now, after completing my undergraduate degree, earning a PhD, and working professionally in the field, I feel deeply grateful for that decision. Engineering has given me more than a technical foundation. It has trained me to think differently, to solve problems creatively, and to approach the world with a mindset grounded in both logic and possibility.

I didn't fully appreciate the power of this training until I observed the contrast between myself and my husband, an environmental scientist who builds predictive models of Earth's climate decades into the future. His work is driven by fundamental principles and long-term simulations, while I tend to think of in terms of constraints, trade-offs, and solutions that must work in the real world today. This contrast isn't just academic. It reveals something essential about engineering: it teaches you to take action, to build, and to deliver even in the face of uncertainty.

ENGINEERING AS A FORCE FOR HUMAN PROGRESS

Engineering has always been a catalyst for human advancement: the steam engine, electricity, the airplane, the semiconductor. All of these technologies began as bold ideas and were brought to life through engineering. Today we are living in another technological revolution: the rise of artificial intelligence (AI) and, more specifically, physical AI.

Unlike traditional AI, which operates primarily in the digital domain, such as processing text, images, and data, physical AI exists in and interacts with the real world. Physical AI perceives its surroundings, reasons, and acts in physical environments. This next wave of intelligent systems is not just about generating answers or recommendations but is also about systems that can drive, build, diagnose, and interact in an intelligent and autonomous manner.

What makes this leap possible is not AI alone but rather AI together with engineering. For centuries, engineering has been the backbone of building the physical world we live in, from transportation systems to medical devices and from power grids to manufacturing plants. Now, engineering serves as the foundation for this next transition, enabling intelligence in systems that interact with the world. Realizing the vision of physical AI demands more than algorithms; it requires the practical ingenuity, systems thinking, and interdisciplinary collaboration that are at the core of engineering.

Electrical engineering provides the sensory and computational infrastructure that powers physical AI. From energy-efficient chips to complex sensor arrays that capture real-time data, electrical engineers design the nervous system of these intelligent machines. At the same time, mechanical engineers make that intelligence mobile and physical. They ensure that these systems can move, grasp, fly, or operate in demanding environments, from robotic arms in factories to drones inspecting power lines in remote terrain.

Computer engineers and computer scientists develop algorithms that enable perception and decision-making. They create neural networks, optimization systems, and learning architectures that allow machines to make sense of their sensory inputs and act accordingly. Civil engineers are applying AI in entirely new ways to reshape the built environment. Imagine smart cities with infrastructure that adapts in real time: bridges embedded with sensors that monitor structural integrity and traffic systems that learn and respond to congestion patterns dynamically.

Together, these disciplines form a powerful network of innovations. Physical AI is not one technology. It is a system of systems, requiring the integration of hardware, software, algorithms, design, and domain expertise. The development of physical AI doesn't call for just better machines. It calls for better engineering.

ENGINEERING THE FUTURE

We are living in an era of accelerated innovation. Technologies that once took decades to develop are now evolving in months. Engineering is not just keeping up; it is at the heart of this transformation. Take autonomous vehicles, for example. They rely on an intricate fusion of electrical systems, sensor networks, real-time computing, mechanical dynamics, and human-centered design. Every piece must work seamlessly in a timely manner to ensure safety, efficiency, and trust. These are not just research prototypes; they are production systems operating in a chaotic, human world.

In the age of physical AI, engineers will be tasked with building intelligent systems that can understand and work in the physical world, systems that are not only smart but also safe, ethical, resilient, and human-centered. These innovations will demand more than technical proficiency; they will require systems thinking, cross-disciplinary collaboration, and a deep sense of responsibility.

Purdue engineering, now with a legacy of over 150 years, exemplifies how institutions can continuously reinvent themselves in response to each wave of innovation. From the industrial age to the digital age and into the era of AI, Purdue has remained a leader by staying grounded in fundamentals while embracing change.

As we look toward the next 50 years, we will undoubtedly see revolutionary shifts in what engineering means and what it enables. Some of today's tools and disciplines may fade, and entirely new ones will emerge. Yet the core of engineering will remain: the problem-solving mindset. By teaching us how to take the unknown and make it work, engineering turns imagination into blueprints and blueprints into reality.

That, to me, is why engineering matters more than ever.

21

ENGINEERING 50 YEARS FROM TODAY

ATLAS

Atlas is a fifth grader in Mrs. Lehman's class.

Mrs. Lehman   Atlas

Engineering 50 years from now

Engineering 50 years from now
will be different from now. We
will probably explore more in space.
We have new devices that can help us
and if we keep doing that, we might
be able to get to space, check if and
after. We might be able to go to
Mars and make a colony.

Life might be easier in the
next 50 years, because we might
have self driving cars everywhere.
We also might be able to turn random
things like dirt into food. And
transportation might be other.

We might be able to solve
global warming, world thirst and hunger.
We might be able to stop or reverse
the effect of global warming. We
might also find a way to filter saltwater
more efficiently. We also might find
a way to make new food.

22

ENGINEERING 50 YEARS FROM TODAY

A Journey of Curiosity, Teamwork, Risk-Taking, and Lifelong Learning

DAVID LI

David Li is the CEO of Ingevity and previously was the CEO of CMC Materials. He has served on board of directors and as an adviser for several companies and was previously a professor at Northwestern University in the Kellogg School of Management.

As a young wide-eyed chemical engineering student at Purdue University over 30 years ago, I embarked on a journey that would shape my career and personal development in profound ways. While my education equipped me with the technical skills to solve complex problems, it also instilled in me even more valuable qualities: the importance of curiosity and adaptability and the confidence to engage with the world. Engineering, for me, was not just about mastering technical knowledge; it was also about cultivating a mindset that embraces challenges, seeks continuous learning, and connects with people from diverse backgrounds. This mindset has shaped not only my career trajectory but also the way I approach life itself. I believe that these qualities will also be valuable to the next generations of engineers in continuing to solve the world's most difficult challenges that the future brings.

THE POWER OF AN ENGINEERING EDUCATION

The true value of an engineering education lies in its ability to build resilience and foster a spirit of exploration. Engineering education gives you the tools to face difficult problems head-on and work collaboratively to find solutions. I still remember early in my career working in a petrochemical refinery when I tried to build a catalyst reactor from scratch and failed spectacularly—both humbling and exhilarating! More importantly, an engineering background empowers you to approach the world with confidence. Engineering isn't just about solving equations; it's about applying that knowledge in real-world contexts, working with teams, and thriving in ambiguity.

Engineering teaches you how to think. It challenges you to analyze problems, formulate solutions, and iterate until you succeed. What stands out most in my journey is how engineering prepared me to approach new challenges with curiosity and determination. Whether working in Shanghai, Seoul, Singapore, Oman, Tapei, or Chicago, I have learned that the most innovative solutions often come from embracing different perspectives and finding common ground. This understanding has been a guiding principle throughout my career.

During my time as an engineering student, I learned that collaboration is just as important as technical skill. Projects weren't solved in isolation; they required teamwork, creativity, and the ability to listen to others' ideas. These lessons became even more critical in my professional life. Engineering doesn't exist in a vacuum—it's connected to people, businesses, and communities. Recognizing the human element behind every technical challenge is what makes engineering a powerful tool whose value will stand the test of time.

TAKING RISKS TO GROW

My career has not followed a conventional path. It has been marked by deliberate, often nontraditional choices to take on opportunities that were sometimes lateral or unpopular but exposed me to new challenges and equipped me with broader experiences. Early in my career, I often took on new roles or assignments without regard to promotion or increased compensation so I could learn a new aspect of the company or business. Whether it was transitioning from process engineering into supply chain or from supply chain into investor relations, each career move represented a calculated risk as I stepped into unfamiliar territories to gain new skills and insights. Stepping into these roles was intimidating at first but ultimately expanded my perspective and prepared me for larger leadership positions. I firmly believe that having the opportunity to take on a new position with the pressure to learn and perform is one of the best positions to be in especially early in one's career, and I encourage the next generation of engineers to take risks (especially early) in their careers.

One of the most pivotal risks I took was moving overseas to manage our Asia Pacific business. Relocating to Asia meant adapting to a new cultural and business environment, building relationships across cultural divides, and learning to lead in a global context. It was a leap into the unknown, but it proved to be one of the most transformative experiences of my life. Immersing myself into different regions and cultures not only sharpened my understanding of global business but also reinforced the importance of flexibility, caring, and open-mindedness in leadership.

As the CEO of CMC Materials, I faced the challenge of leading a company to regain its growth after a prolonged period of stagnation. The business had lost market share, and growth had stalled. To rejuvenate the organization, I knew that I had to take bold risks and make changes quickly. These included reorganizing the executive team, pursuing acquisitions, and driving a cultural transformation. Changing a company's culture is no small feat—it requires clear vision, consistent communication, and a willingness to challenge the status quo. But these efforts paid off. During my tenure we outpaced industry growth, delivered significant return to shareholders, and ultimately sold the company for a valuation that exceeded even our own lofty expectations.

A LIFELONG JOURNEY OF
CURIOSITY AND LEARNING

Curiosity is at the heart of engineering and is also a quality that drives lifelong learning. Throughout my career, I have continually been drawn to new challenges—whether it was leading a semiconductor materials company to exponential growth or advising firms on their most complex strategic decisions. This innate curiosity has kept me moving forward, always eager to learn more, explore new opportunities, and adapt to changing markets.

I never expected to be a CEO, to work in semiconductors, or to stand in front of a classroom as a professor in Northwestern University's Kellogg School of Management. Yet having the curiosity to learn and the willingness to take on challenges has given me the journey of a lifetime. Each unexpected step has taught me invaluable lessons and opened doors to opportunities I couldn't have imagined.

In the coming decades, the importance of being a lifelong learner will only grow. As we look to the future of engineering, we see exciting developments on the horizon—fields such as artificial intelligence, life sciences, quantum computing, and renewable energy will likely shape the next era of human progress. These innovations promise to redefine industries, create new opportunities, and solve problems we can only imagine today. But to stay at the forefront of these advancements, engineers must be committed to constant learning and growth.

EMBRACING THE UNKNOWN

Taking risks also means embracing uncertainty and understanding that failure is a stepping stone to success. I have always believed that great careers are built not just on achievements but also on the lessons learned from taking informed risks. During my tenure at CMC Materials, for example, we made two major acquisitions to expand our market presence. These were not guaranteed successes—they required strategic planning, cultural integration, and a willingness to navigate the unexpected. Yet these bold moves ultimately transformed the company and positioned it for long-term growth.

Just as I have evolved throughout my career, today's students must be prepared for a future that will look very different from the present. Engineering 50 years from now will likely involve technologies and challenges that we can't yet foresee. However, what will remain constant is the need for engineers to think critically, remain curious, and approach every problem with the drive to learn and innovate.

A SHARED PURPOSE FOR THE FUTURE

As we look to the future of engineering, I encourage today's students and future generations to embrace the unknown. You are entering a world of unprecedented opportunities where technology, collaboration, and creativity will come together to solve the world's biggest challenges. From sustainable energy to breakthroughs in health care, the potential impact of engineers has never been greater.

It is also important to remember that the journey doesn't end with earning your degree. Be open to new experiences, whether that's diving into a different field, leading a team, or taking on an entirely new role. My own career has taken me across industries and continents, from technology and energy to academia and philanthropy. Each step of the way, the principles I learned as an engineer—problem solving, teamwork, informed risk taking, and curiosity—have served me well.

Finally, I hope that as we move forward we can find more ways to collaborate across differences—whether of culture, geography, or perspective—and work together to build a better future. In a world that sometimes feels increasingly divided, engineering can remind us of the power of common ground and shared purpose.

GRATITUDE AND ENCOURAGEMENT

None of this would have been possible without the support of those who believed in me: my family, friends, mentors, and colleagues. I especially owe a debt of gratitude to my parents, Drs. Norman and Jane Li, for instilling in me the value of education and lifelong learning, and to my family including my wife Sophia and my children Ethan, Henry, and Olivia for always believing in me. Their encouragement has been my greatest source of strength.

As I reflect on my journey, I am filled with gratitude for the opportunities I've had and the people who helped me along the way. And to the next generation of engineers, I encourage you to seize every opportunity, stay curious, take calculated risks, and never stop learning. The world is full of challenges waiting for solutions, and I am confident that engineers will be the ones to solve them.

23

THE SPACE RUSH

A Rush for Resilient and Sustainable Habitation and Industrialization on Earth and in Space

AJAY P. MALSHE

Ajay P. Malshe is a Purdue Goodson Distinguished Professor and an entrepreneur. He copioneered award-winning factories-in-space for servicing, assembly, and manufacturing concept. Malshe holds 27+ patents, has contributed to 225+ publications, and has received 45+ global recognitions in advanced manufacturing, materials, and design.

America has seen many a rush in recent centuries, including the gold rush in the mid-nineteenth century, dotcom in the late twentieth century, and now space. A rush is a phenomenon that engages human curiosity for new knowledge and passion for continuous advancement, with industrious humans and industrial machines in the loop for the betterment of individuals and society. The following is a critical outlook for human coexistence and operation on Earth and in space, and industrialization is in that pursuit.

WHY SPACE FOR THE NEXT
INDUSTRIAL REVOLUTION?

Humanity stands at the cusp of the next great industrial revolution, driven by advances in artificial intelligence (AI), robotics, 3D printing, hybrid manufacturing, biotechnology, materials, designs, energy systems, and so on and new business models. Yet the most transformative arena for the fourth industrial revolution is not on Earth—it is in space. Space presents unique conditions, including microgravity, extreme temperatures, vacuum, and abundant resources such as rare minerals and solar energy, which could enable groundbreaking innovations that are impossible on Earth.

Manufacturing in microgravity can create materials with superior properties, such as flawless materials, advanced pharmaceuticals, next-generation semiconductors, and new manufacturing processes, products, and services. Similarly, the vast untapped reservoirs of helium-3 and rare earth metals on the moon and asteroids could meet Earth's increasing demand for sustainable energy and critical materials.

Moreover, the logistical constraints of working in space naturally drive efficiency, miniaturization, and technological synergy—key enablers of innovation. In this sense, space becomes not just a frontier but a laboratory for rethinking industrial processes, creating self-sustaining ecosystems, and fostering collaboration between nations and sectors. Space is also the

Five predictions from the space rush for key events in space in the next five decades:

1. The *first child born* in space
2. The *first human marriage in space attended by both mates* in space
3. The birth of the *first space currency*
4. The *first international conflict*
5. The *first habitat opened on the moon* for extended living

next frontier for defense in a multidomain landscape. Defense and commerce are drivers for the new space rush!

WHY IN-SPACE DEFENSE AND COMMERCE GO HAND IN HAND

The interdependence of defense and commerce in space is a reflection of its dual-use nature. Many technologies developed for space exploration, such as satellite communication, remote sensing, and propulsion systems, serve both military and commercial purposes. For instance, satellite constellations support military operations, provide global internet coverage, and enable financial transactions.

As nations seek to protect their assets in orbit—be it satellites, space stations, or resource extraction facilities—defense and commerce naturally converge. A robust defense infrastructure ensures the security of commercial investments, while private-sector innovations drive down costs and accelerate advancements in space technologies.

Furthermore, geopolitical competition in space is becoming a reality, with nations such as the United States, China, and Russia making significant investments in space defense. A well-coordinated partnership between defense and commerce can ensure that space remains a zone of sustainable growth rather than conflict.

WHAT UNIFIED IN-SPACE AND ON-EARTH EDUCATION PLATFORMS LOOK LIKE

Education will be the cornerstone of a thriving space economy and society. A unified education platform should seamlessly integrate terrestrial and in-space learning, focusing on STEM (science, technology, engineering, mathematics) while fostering interdisciplinary collaboration.

Virtual reality and augmented reality platforms could allow students on Earth to virtually explore space habitats, operate robotic arms, and conduct experiments in microgravity. Similarly, astronauts and researchers in orbit could collaborate with students and educators on Earth in real time, sharing data and insights to enrich learning experiences.

This platform should also emphasize accessibility to all forms of identities. By leveraging satellite connectivity and open-access resources, even remote and underserved communities can participate in the space economy. In doing so, the platform ensures that space becomes a shared human endeavor rather than an exclusive domain.

WHY DIGITAL TECHNOLOGY IN SPACE?

Digital assets (semiconductors, electronics, digital technologies, etc.) and physical assets are cornerstones of modern commerce and defense, enabling everything from smartphones to AI to satellites. Producing digital assets in space could unlock new levels of performance and efficiency and operate both on Earth and in space.

New products and services include data servers, quantum manufacturing, cross–Kármán line digital twins, and AI agents empowering digital technologies in space for Earth and space. For example, in microgravity, materials crystallize more perfectly, creating semiconductors with "perfect," almost defects-free materials. Devices and chips manufactured from such "perfect" materials could enable quantum devices, faster processing speeds, lower energy consumption, and enhanced durability—critical for space- and Earth-based applications (e.g., quantum computing). Also, space offers a "free" vacuum desired for semiconductors and other manufacturing processes.

The demand for chips is skyrocketing, driven by the proliferation of Internet of Things devices, 5G networks, and AI applications. Establishing semiconductor manufacturing facilities in orbit could alleviate Earth-based supply chain bottlenecks, reduce geopolitical risks, and ensure a steady supply of high-performance chips for future technologies.

WHAT ARE FIVE LOW-HANGING PRODUCTS IN SPACE?

There are five long-hanging products for Earth:

1. *Pharmaceuticals*: Proteins and crystals grow more uniformly in microgravity, enabling the production of more effective drugs.
2. *Fiber optics*: ZBLAN fiber optics manufactured in space offer unparalleled signal clarity and transmission efficiency.
3. *Space tourism*: As costs decrease, suborbital and orbital tourism will become more accessible, driving demand for luxury space experiences.
4. *Earth observation data*: Satellites provide critical data for agriculture, disaster management, and climate monitoring.
5. *Solar power beaming*: Space-based solar panels could transmit clean energy to Earth via microwave or laser technology.

And there are five long-hanging products for space:

1. *Propellant production*: Harvesting water from the moon or asteroids to produce hydrogen and oxygen for rocket fuel.
2. *3D-printed components*: On-demand manufacturing of tools and replacement parts in space habitats and missions.
3. *Radiation-resistant materials*: Developing advanced materials to protect astronauts and equipment from cosmic radiation.
4. *Oxygen and water recycling*: Closed-loop life-support systems to sustain long-term missions and habitats.
5. *Habitat modules*: Modular construction of space habitats for research, tourism, and colonization.

WHAT ARE THE POLICY BARRIERS FOR SPACE COMMERCE AND DEFENSE?

Space commerce and defense face significant policy challenges, including

1. *Ambiguities in space law*: The Outer Space Treaty of 1967, while foundational, lacks clarity on issues such as resource ownership and military activity in space.
2. *Export control regulations*: Strict national controls, such as the US International Traffic in Arms Regulations, hinder international collaboration and technology sharing.
3. *Liability and insurance*: Determining liability for orbital collisions and debris-related damage remains unresolved.
4. *Spectrum allocation*: The competition for radio frequencies among satellite operators, governments, and terrestrial networks poses regulatory hurdles.
5. *Equitable access*: Developing nations and private entities should have fair access to space resources and opportunities.

WHAT WILL INCLUSIVE SPACE WILL BE LIKE FOR ALL HUMANKIND AND OTHER SPECIES

A universal space represents humanity's collective interests. Symbiosis in space could mean creating opportunities for all individuals and nations to participate in exploration and commerce. It could also involve addressing ethical questions, such as the rights of other species and the preservation of extraterrestrial environments.

For example, long-term space habitats may need to support biodiversity, including plants, animals, and pets, to create sustainable ecosystems. Similarly, protocols must be established to prevent the contamination of celestial bodies with Earth-based microbes.

By fostering collaboration, a universal space can serve as a unifying platform to tackle humanity's most significant challenges, from resource scarcity to climate change.

CONCLUSION

The industrial revolution in space is not a distant dream—it is unfolding now. By embracing space as the next frontier for innovation, defense, commerce, and education, humanity can unlock new opportunities for growth and collaboration.

However, realizing this vision requires addressing policy barriers, fostering inclusivity, and ensuring that the benefits of space exploration are shared equitably. With careful planning and cooperation, space can become a platform for a brighter, more sustainable future for all humankind and perhaps for other species as well.

24

ENGINEERING
50 YEARS FROM TODAY

KAMBRIE

Kambrie is a fifth grader in Mrs. Lehman's class.

Engineering 50 years from today Mrs. Lehman

I think there would be more AI will be used a lot more then it is used today. like fields will have AI and machine learning, renewable, aerospace engineering, and cybersecurity. And there will more small homes. Computers are likely to be significantly smaller and more powerful, with processing speeds potentially reaching petahertz levels. That is what I think engineering will be like in 50 years.

I think Development of the internet, Smart phones, and global positioning system will be better. I think there will also be immersive virtual and augmented Realities. I think there will be advanced AI in areas of healthcare, energy, and transportation. I think in 50 years there will Airports for flying taxis. that is what I think there will be engineered in 50 years.

25

AN INCOMPLETE STORY OF ENGINEERING PAST, PRESENT, AND FUTURE

KIERNAN MCCULLOUGH

Kiernan McCullough earned his bachelor's degree in nuclear engineering at Purdue University and his master's degree in radiological medical physics at the University of Kentucky. He is currently the chief of therapy physics at Colorado Associates in Medical Physics and specializes in therapeutic medical physics.

I wish to preface this essay with an admission to the reader. The absurdity of predicting how engineering will be shaped 50 years from now is not lost on me, and I have little clout to support my assertions. The explosion of artificial intelligence (AI) technology and its recent prevalence could not have been better timed, for I would have neither the knowledge nor the resources without it to construct this essay. Finally, I greatly appreciate this opportunity to contribute to the celebration of the remarkable history of Purdue engineering, and my opinions herein are heavily biased by a four-year stint in the Nuclear Reactor Lab in the depths of the Electrical Engineering building.

In preparation for this project, I thought it prudent to peruse trends in engineering, novel breakthroughs over human history, what drove engineering during that time, and how it was practiced (thanks again, to AI). While this did not conjure a crystalline image of what the future holds as I may have hoped, it was inspiring and thought-provoking to see the massive advances in science and technology that have marked our civilization. Incredible achievements have shaped the way we move, eat, communicate, fight, heal, and think. Moreover, it was elucidating to see the incredible breadth of human engineering and the ever-changing focus of technological efforts to solve the newest and most pressing issues.

I chose at least to narrow my scope to engineering (mostly) in the United States, starting 150 years ago: the birth of engineering at Purdue University and the impetus for this assignment. This was the post–Civil War Reconstruction era, where the US Army Corp of Engineers was tasked with reshaping the country's infrastructure and explore the American West. Heavy investment in railroad expansion and the telegraph dominated engineering efforts to better connect and repair the war-torn country. This was facilitated by the rapid expansion of industrialization and manufacturing advances, with major funding sources being the government and the industrialists who drove the expansion of transportation and communication. Following Reconstruction, the Gilded Age from the late 1870s to the late 1890s, and the Progressive Era of the late 1890s to the 1920s that saw a major boon in private wealth and innovation, fueled by the US entry into World War I, we were gifted with the airplane, electricity, the Model T, the tank, and the advent of consumer electronics. Within the physics realm, Albert Einstein's annus mirabilis changed our understanding of reality in 1905 with the publication of four revolutionary papers. The general growth in prosperity during this time and into the Roaring Twenties allowed for increased private investment into the scientific realm, with much focus on consumer products and entertainment. Simultaneously, the quantum world was beginning to unravel, with Niels Bohr winning the Nobel Prize in Physics for his work on quantum theory in 1922.

The rise of the Great Depression in the 1930s saw a return to primarily government-funded projects, with interest in technology spurred by the start of World War II. The pressure cooker of the 1940s led to some of the most rapid technological advancements in history: jet rocketry, early computers, radar, and the nuclear bomb, with each invention playing a pivotal role in redefining the rules of engagement. Large-scale engineering projects up to this point had primarily been composed of laying steel and running wire, but now one of the largest expenditures in American history was devoted to making theoretical physics practical: cracking the atom. For better or worse, the world at war shined an unprecedented light on the engineering and physics world. Great minds were treasured above all else, with both sides eagerly awaiting solutions to the same mystery. I believe that during this period, the scope of engineering practice and the role it played in society began to pivot. Advancements in nuclear physics, quantum mechanics, aerospace technology, and computer science all broke open humanity's understanding of the rules of the universe. While it may take many decades for these breakthroughs to permeate the masses, the discoveries of this period permanently changed the role that technology will play in human history.

Government-fueled research continued to be the predominant source of investment well after World War II and into the Cold War. Aerospace technology and the Space Race dominated the era, with computation improving in lockstep to culminate in Purdue's Apollo 11 mission. In the medical world, James Watson and Francis Crick won the Nobel Prize for their discovery of DNA in 1962, rewriting our understanding of biology and setting the stages for genetic engineering and a revolution in personalized medicine. The advancements of the 1950s and 1960s provided a ripe environment for private investment and major corporations to enter the research arena. In the 1970s we entered the information age, a decades-long focus on computer technology, electronic communication, and the ever-shrinking world of transistors. Continuous refinement of electronic technology allowed for an explosion of personal devices: home computers, cellular phones, and the rise of the internet introduced immediate access to information and instantaneous communication.

Again, I believe that a paradigm shift began to emerge. Venture capital groups and tech giants begin to shape the social landscape and direct technological investment that is driven more by consumerism than government interest. In turn, dependence on electronics and technology began to evolve. Until the 1990s and 2000s, engineering and tech advancements primarily helped us to either work or play, but now technology offers newfound capacities to help us be. Social media emerged, giving individuals a unique opportunity to shape a subset of their lives; the online avatar blends humanity and technology in a way that augments our sense of self and allows people to live more and more of their lives via the internet. In this period, personal data and habits became a form of currency; internet services and social media giants are some of the most powerful companies on the planet.

Advancements in engineering are abundant as we speed toward the present. Increased accessibility to powerful computing resources and custom 3D fabrication allows for smaller teams to tackle ambitious projects backed by a growing number of venture capital groups willing to gamble on the next big breakthrough. As a result, engineering feats once only achievable through massive government coordination are increasingly shifted to the private sector: space travel, self-driving cars, quantum computing, blockchain technology, and AI, to name a few. We are again at the brink of another revolution it seems, with some dubbing this moment as the transition from the information age to the intelligence age: a noticeable shift where computers are no longer helping us think more efficiently but instead are "thinking" for us. In particular, the advent of commercial AI solutions has had an immediate and dramatic impact, allowing for more efficient and equitable access to research and even the capacity to form new hypotheses for humans to scrutinize rather than incept. The capacity to adopt these tools into powerful creative engines is already a determining factor in who has the advantage.

For an essay describing engineering in the next 50 years, I appreciate that I spent an awful lot of time reflecting on the past. However, the lessons baked into this historical review may help shape our picture of the future. While the power and resources to inspire change were once reserved for the government, the rise of megacorporations and the push toward globalization have allowed many players to enter the sphere, with few tethers to ground their initiatives. We also see a cyclical nature of progress: major breakthroughs dot our history every 30 years or so, followed by a period of refinement, adaptation, and commercialization that then begets another pivotal change. The difference we see over time, however, is that these secrets were once heavily guarded by government enterprise but are now more public and capable of adoption than ever before. As such, I would imagine that the timelines between periods of technological upheaval will condense as the capacity to digest and implement new research methods accelerates. This of course doesn't take into consideration the unforeseen challenges ahead that may force our collective hand toward a single goal, either together to avoid catastrophe or apart as we have already witnessed in global conflict.

Perhaps another leap is due. We are now 30 years removed from the internet revolution, and the socialization of AI feels like the spark that will ignite the next flame. In the coming decades, AI tools (perhaps "teammates" even) that are fine-tuned for each task will be vital members of the research environment. Hypothesis vetting, prototyping, and virtual simulation, all guided by AI and fueled by quantum computers, will allow for rapid progression and industrialization of advances to come. Renewable resources and high-density sources of energy will be of paramount importance, as the demands of our globalized tech-heavy industries will only increase as the classical sources of fuel are repleted. The impact of global climate change may prove to be a unifying obstacle that encourages humanity to collaborate on an epic scale to arm ourselves against the growing threat of inclement weather and the unknown consequences of disintegrating polar ice. This will also affect the evident push toward space colonization, with lunar and Martian occupation currently being strategized, even going so far as the Nokia Corporation already planning for cell phone coverage on the moon!

I estimate that in the next 30 years, the intelligence revolution we are currently experiencing will ultimately culminate in two major changes that will set the stage for the next great leap forward. First, there will be destabilization of the current work environment as we establish a new normal alongside rapid advancements in AI, robotic replacements, and automation. How these changes impact our society will ultimately be the responsibility of engineers in terms of building quality assurance tools and cultivating this technology for positive change. Whether good, bad, or ugly, the roles that humans and technology play will begin to stabilize as the world settles into a larger global economy that is more focused on resource efficiency and funded by megaconglomerates. Second, decades' worth of technological breakthroughs, accelerated by AI, will accumulate such that the most challenging engineering problems of our history may finally be achievable. Solving the riddles to atomic-level fabrication, controlled nuclear fusion, genetic medicine, interplanetary space travel, and a Purdue national basketball championship would mark the next monumental leap forward in humankind. With access to an infinite energy well, trusty robotic companions, and budget space travel, humanity truly has the capacity to dive into the future that Arthur Clarke predicted in *2001: A Space Odyssey*.

Come 2075, the new resources at our disposal will cause another disruption and shuffling of financial powers as we have seen time and again. At this point, technology will have the capacity to integrate with the human; to what degree we meld will probably be driven as much by the consumer as the military, with brain-computer interfaces and bionic implants revealing a new realm of predictive medicine, thought transfer, collaboration, and combat strategy. Humans will have to determine what roles we protect for ourselves as intelligent systems become capable of most human tasks. At this point we will likely be punished or rewarded for our actions through the 2050s. Did we galvanize as a species to prioritize responsible resource allocation and thoughtful living? Have we been myopic and chased short-term riches at our own peril?

If there is something I have learned through this exercise, it is that engineers have been the central cog that moves us from the past to the future. Engineers are the ones who don't see problems, just solutions waiting to be uncovered. While preparing for this essay, I stumbled across a quote from Billy Vaughn Koen's *Definition of the Engineering Method* that exemplifies why engineering was and will always be a societal driving force: "the engineering method . . . [is the] strategy for causing the best change in a poorly understood or uncertain situation within the available resources."[1] Simply put, the engineering method is the recipe for success and is applicable no matter what problem, what industry, or what decade in which you may face it. Thanks to this tenet, I am confident that we as Boilermakers will be living on the bleeding edge, prepared for the uncertainties of the future, thanks to the lessons imparted by our 150-year past.

Hail Purdue!

NOTE

1 B. V. Koen, *Definition of the engineering method* (American Society for Engineering Education, 1985), 10, https://scispace.com/pdf /definition-of-the-engineering-method-4n9pzpd3p1.pdf.

26

THE SECOND LAW OF GEOPOLITICS

COLIN MICHAEL MCGONAGILL

Colin Michael McGonagill is the chief executive officer of McGonagill & Bay Exploration, a Texas-based energy company producing primarily from the Permian Basin.

Engineering has always preceded geopolitical consequences. The term "engineer" dates back to 1325 as the "constructor of a military machine," usually a clever individual giving range to warfare with no more than wood and some twine (Oxford University Press, n.d.). Deepwater navigation turned a flat world round, forever proving Pythagoras and Aristotle right. Nicolas Leonard Sadi Carnot, born in 1792, was the son of the minister of war under Napolean and was every bit a mathematical Carl von Clausewitz. Carnot understood that the center of power in his modern era would be whoever would control the steam engine. He realized, brilliantly so, that nature had a dissymmetry and that whoever could move among that dissymmetry the most efficiently would control it. He understood that the world moved from a state of order to a state of chaos. This generalization became the second law of thermodynamics stated by Lord Kelvin in 1851.

The application of the steam engine in the 19th century gave birth to the first boilermakers, thermodynamic cowboys who risked their lives to capture temperature and pressure. This now obtained energy propelled the world into the industrial revolution and was accelerated by the likes of Nicolaus Otto and Rudolf Diesel.

No other part of science has contributed as much to the liberation of the human spirit as the Second Law of thermodynamics. Yet, at the same time, few other parts of science are held to be so recondite. . . . Not many would pass C. P. Snow's general test of general literacy, in which not knowing the Second Law is equivalent to not having read a work of Shakespeare. (Atkins, 1984, p. vii)

My thesis is that the focus of engineering 50 years from now will follow the second law of thermodynamics, which will be most found at the center of gravity of whatever geopolitical events are of the age. Why does geopolitics matter? It matters because geopolitics dictates what engineering problems are prioritized and funded; thus, predicting these events becomes a crucial part of forecasting engineering. The framework I use to predict these events is quite simple:

1. Humans do not choose where they are born, and
2. All humans eventually die.

While sounding morbid, these two rules underpin most strategic decisions at the nation-state level. How many wars have been started to create a new border, and how many empires no longer exist today?

Population is one variable about which there is great certainty—the adults of tomorrow are children today. Thus, the conclusions reached are locked in, and the consequences fall into the category of the known/ known, to borrow from Donald Rumsfeld. This fundamental fact must be appreciated by the national security community, whether they are scholars or policy analysts. They do have an excellent window into the future and thus the luxury of time to prevent some, at least, of these problems. (Thayer, 2009, p. 3090)

This pseudo "second" law that "all humans eventually die," overlayed with geography, is the single most important tool I use to help predict the outcome of energy markets, technology, and engineering. It has proven to be *the* tool that energy traders and CEOs must understand. Here are some hard facts (and questions) that anyone who has looked at a map and watched the news will understand:

1. Russia has the largest land border and also has an inverted population pyramid and is not reproducing at a sufficient rate to sustain its population. How do you defend the largest land border with a shrinking population that will eventually approach zero given enough time (Drelichman & Voth, 2024)? What happens to Russia's natural resources and nuclear arsenal when there are no more people to manage them? Unfortunately, I doubt that ChatGPT has an answer to this question. Pulitzer Prize winner Daniel Yergin understood the second law of geopolitics and accurately predicted the current situation in *Russia 2010*, coauthored with Thane Gustafson and published in 1993. I think this proves that humans will always be in the loop at some point in the thinking process.

2. China accounts for 35% of worldwide manufacturing capability (Drelichman & Voth, 2024) and has an inverted population pyramid that is worse than that of Russia. Fifty years from now China's population will be roughly half of what it is today and after twenty-five years more only a third of what it is today (PopulationPyramid. net, 2024). How do you build the same amount of stuff for the same price in that environment? How do you maintain relevance, especially when your neighbor, whom you have been in a civil war with for almost 100 years, makes one of the most relevant devices (computer chips) of the modern world? The US Department of Defense (2024) and the US Air Force (Reich, 2024) both have jointly warned that this issue is coming to a head at a very rapid rate. Purdue's own funding from the CHIPS Act and the Maurice J. Zucrow Laboratories hypersonic research facility affirm this reality.

Fifty years from now, "Made in China" will not be the norm due to either population collapse or embargos. I often tell my wife to imagine a world where you go to Target or Walmart and there is almost nothing on the shelves. What would you do? For one, you would start making things. Manufacturing is going to make a big comeback, and the advent of additive manufacturing is going to make prior impossible designs possible. Note: the first draft of this essay was written before President Donald Trump's tariffs announcement on April 2, 2025, accelerating this timeline.

The fall of the Russian Federation is going to be a big nuclear problem. Fortunately, the world has had this happen before with the Soviet Union. The megatons to megawatts program in 1993 is hailed as the most successful disarmament program ever conceived. Most Americans do not realize that between 1993 and 2013 almost 10% of electricity usage in the United States came from Russian bomb-grade nuclear material (Centrus Energy, n.d.). I believe that we will see a resurgence of this program, with small modular nuclear reactors collocated with data centers as the inevitable outlet for this material. The key will be bargaining it out of the hands of oligarchs and tyrants; I pray that God gives us skilled negotiators.

This sets the stage for answering the actual question of where engineering will be 50 years from now. Remember, I said that the second law of thermodynamics will be found at the center of the second law of geopolitics. I predicted above a resurgence in manufacturing capability and nuclear power generation at the core of what we will do. What are some secondary spin-offs? I think there will be a resurgence of power cycle and turbo machinery design. Most if not all of the existing machines were built without additive manufacturing in mind. I love watching Formula One, and I am amazed at how much power and efficiency these racing cars have due to additive-based designs. Once the larger manufacturers start to figure out how to build engines at scale using this technology, I think we could see some new engines, similar to how Otto and Diesel converted external combustion to internal combustion. The key will be building the machines that build the machines, and that won't happen until there is an extreme need (i.e., capital influx) for new manufacturing to be built (i.e., China-Taiwan catalyst).

The advent of large language models has proven that it is possible to take a billion-parameter problem and condense it down to meaningful words. I believe that the infinite parameter nature (upstream requirement for the needed compute) lends itself to more specific applications. I think that the largest application of the above will be in genetic engineering. For instance, a single gram of DNA can hold 215 petabytes of data. If anything, genetic material would make a great hard drive, but I think that being able to selectively program DNA will become a center of power that people will crave and want. Think about being able to "program" yourself to be smart, to have a certain hair color, or to have immunity to diseases. There will be inevitable ethical issues with this, of course, but if the COVID-19 mRNA vaccine taught me anything, it is that DNA and RNA are now fair game to be tinkered with at scale. Think CRISPR meets ChatGPT. I believe that there will be ways to store data inherently in yourself and to alter your existing data, which could give you an edge or augment life in new meaningful ways. I think that additive manufacturing will be required to build the tools that will be needed to build the DNA. I am generally not an advocate for transhumanism, but I think that the tools are there for it to be discovered.

So, what about oil and gas? That is my day job, after all. Liquefied natural gas (LNG) will be the diplomatic molecule, like the petrodollar of the 1970s. The cryogenic nature of LNG gives it an edge over oil in certain applications. LNG will be the preferred fuel for spacecraft and eventually aircraft, especially with pressure gain combustion on the rise. Most people do not realize that the United States only has 20–30 days of crude oil supply on hand at any point in time. Conflicts generated by the two geopolitical certainties could consume this, and we will have to make more. Russian crude will go offline at some point unless China moves up from its northern border to turn it on for itself. The rank order of commodities in the next 50 years will be oil and then, in order, copper, LNG, and coal, all utilized to make sure the Russian uranium/plutonium problem and the China-Tawain problems do not get out of hand. Rare earth minerals are cool, but how are those mining rigs powered? I haven't seen a nuclear-powered dump truck yet, and I doubt that the US Nuclear Regulatory Commission will approve one.

Let's bring this back to Carnot, the original inventor of the second law of thermodynamics. The world will always move from a state of order to a state of chaos. Being able to efficiently move in that world is the key. To phrase it more geopolitically, everyone knows that life isn't fair, but those who can make life fair have power. Bretton Woods proved this, and the new global order won't be much different. The interchange of thermodynamics and geopolitics is indispensable. We can spend our resources fighting in wars or spend our time preventing them. The scenes will be similar, but the outcomes will be quite disparate. The inventors of the art knew this. You should too.

REFERENCES

Atkins, P. W. (1984). *The second law: Energy, chaos, and form*. Scientific American Books.

Centrus Energy. (n.d.). *Megatons to megawatts: A history of nuclear security and nonproliferation*. https://www.centrusenergy.com/who-we-are/history/megatons-to-megawatts/

Drelichman, M., & Voth, H.-J. (2024). *China: The world's sole manufacturing superpower? A line sketch of the rise*. VOX EU—Centre for Economic Policy Research. https://cepr.org/voxeu/columns/china-worlds-sole-manufacturing-superpower-line-sketch-rise

Oxford University Press. (n.d.). *Engineering*. Oxford English Dictionary. https://www.oed.com

PopulationPyramid.net. (2024). *China population pyramid 2024*. https://www.populationpyramid.net/china/2024/

Reich, S. (2024). The ambitious dragon: Beijing's calculus for invading Taiwan by 2030. *Journal of Indo-Pacific Affairs, 6*(3) (March–April), 37–53. https://www.airuniversity.af.edu/JIPA/Display/Article/3371474/the-ambitious-dragon-beijings-calculus-for-invading-taiwan-by-2030/

Thayer, B. A. (2009). Considering population and war: A critical and neglected aspect of conflict studies. *Philosophical Transactions of the Royal Society B, 364*(1532), 3081–3092. doi:10.1098/rstb.2009.0151.

US Department of Defense. (2024). *Military and security developments involving the People's Republic of China 2024.* https://media.defense .gov/2024/Dec/18/2003615520/-1/-1/0/MILITARY-AND-SECURITY -DEVELOPMENTS-INVOLVING-THE-PEOPLES-REPUBLIC -OF-CHINA-2024.PDF

Yergin, D., & Gustafson, T. (1993). *Russia 2010: And what it means for the world.* Random House.

27

ENGINEERING 50 YEARS FROM TODAY

BRENTLY

Brently is a fifth grader in Mrs. Lehman's class.

Brently #9   mrs.Lehman

Engineering 50 years from today

What will it be Like in 50 years?
I think there will be Flying cars.
50 years is a Long time for a Lot of things to happen.
In 50 years there are Probably Robots.
there IS a Lot going to be here in 50 years.

In 50 years I think there will be Holograms
on iPhones and apple watches and iPads.
there is gonna be AI. Russia will change
Alot. Prosthetics will Be different In alot of
ways. there is going to be Medicine that makes
u healthier and makes yo live longer

In 50 years there will robots that can
Reduce Loniliness For humans. Also Food will increase
for us, to saving space. It can be a good thing for us.
Having all these resorses For us.

28

THE CONSTANTS OF PURDUE ENGINEERING

ABIGAIL MIZZI

Abigail Mizzi earned her BS in aeronautical and astronautical engineering in 2025. She was the 2024–2025 president of the Purdue Society of Women Engineers, Purdue Space Program propulsion lead, and College of Engineering ambassador. With a master's degree research focused on fluid dynamics in micro-gravity, she will also fly on Purdue 1.

Fifty years from today, there will be more Purdue engineers in the world than ever before. They will be the most skilled, hardworking, dedicated, and dependable engineers around. They will come from all varieties of backgrounds across the nation and the world and will use their education and skills to break boundaries and create innovations that allow humankind to take the next giant leaps. Purdue engineers will bridge boundaries, connect people, encourage positive change, and design better technology to integrate within our world. From my past four years here at Purdue, I can say with certainty that excellence will continue to be taught and strong engineers will continue graduating from this university no matter how many decades go by. This is because Purdue engineers lift each other up and support each other through the highs and lows of engineering. As in any good engineering problem, these are the constants of Purdue engineering that will remain no matter the year or decade, as they are the foundation of Purdue engineering and the governing laws that make the institution so strong.

Therefore, 50 years from today, the friends I have met and made in my time at Purdue will have grown to become truly impactful and inspirational leaders of the world. They will continue the Purdue legacy of innovating and leading in incredible ways across all disciplines of engineering, from supply chains, fighter jets, and electric propulsion to nuclear power. My fellow classmates and friends will have risen to be PhDs and presidents and will lead companies and make decisions that make impacts locally, nationally, and globally. I have seen this already during my years at Purdue. In reaching out to and asking for advice from Society of Women Engineers (SWE) board members from 10 to 20 years ago, I have received kindness and the commitment to support others within the Purdue community. During the four years that my classmates and I have been here, we have been tasked with hard challenges, determining how to best use resources as student leaders growing organizations and membership bodies, collaborating between organizations, and gaining audiences with corporate vice presidents, CEOs, politicians, inventors, and news reporters. No matter what stage Purdue engineers take, they find the areas of improvement, market to the crowds, fill technology gaps, voice reason, and lead decision-making councils.

As the current president of Purdue's collegiate section of the SWE, I have seen how this organization acts as a microcosm of Purdue engineering and demonstrates the strength and leadership potential of Purdue students. I am honored to lead the Purdue SWE section, as we are the longest continuously chartered SWE section in the country and one of the largest collegiate sections nationwide. Our section serves over 600 members and is led by a board of 45 student leaders who organize professional development events and social gatherings, lead four unique technical teams, and host K–12 STEM (science, technology, engineering, mathematics) outreach programs year-round.

This leadership potential is grown through working hard on engineering each year at Purdue, which has allowed Purdue graduates to persevere through tough days personally and professionally. Looking at graduation soon and reflecting on an engineering degree has taught me to be resourceful, ask for help and realize my limitations, break down a problem and understand my assumptions and personal biases, and utilize the tools to my advantage to solve a problem creatively. These are all skills transferable to wherever I go after graduation in the business world in terms of leading teams, building aircraft, and designing rocket engines. I know that these skills will guide me for the rest of my career and propel to be a strong technical and professional leader. This aligns strongly with my life goals to work hard on valiant causes and use my efforts to better the world, which is possible with an engineering degree and the work ethic and problem-solving skills taught to me by Purdue engineering staff and the people I have met here throughout my four years. My time at Purdue went by fast and taught me a lot about not only aerospace engineering but also the people whom I hope to continue surrounding myself with in the future, how to ask for help when life is tough, and how to be my best self as a teammate. I care about people and want to support my teammates to be their best selves because we are stronger together, and that is what engineering is: bridging skills and knowledge to soar higher together. Every day as I walk through the Neil Armstrong Hall of Engineering, I am grateful for the people whom Purdue engineering has brought into my life because they are some of the best. And I am excited to see how we use our knowledge to make all our goals reality.

So, although I can't predict the future, there are certain constants that can be assumed and are certain in regard to Purdue engineering. The engineers who leave this institution can be found in every company across disciplines using their engineering skills to solve complex problems, encourage change, and connect the world. One example was when a past recipient of a scholarship I received reached out to me and congratulated me on the scholarship I won. Now in his career three decades after graduating from Purdue, he is the vice president of a large aerospace manufacturer and told me anecdotes about running into his friends from undergraduate school while on international travel to meet with a large aircraft manufacturer. When he walked into a meeting, he saw one of his friends who is a vice president of the company to which his own company supplies aircraft parts. From coast to coast when I was on a visit to the NASA Langley Research Center, the director of the test facilities had a Purdue connection due to working alongside my research professor in high school. Another instance of the Purdue community transcending distance was when I worked in southern California for a summer internship; my coworker was pursuing her master's degree from Purdue. These types of interactions among Purdue engineers have occurred across the country in happenstance meetings, online connections, meeting Purdue alumni at the SWE 2024 national conference in Los Angeles, and as a prospective student in high school meeting the current Purdue SWE president, who offered me her phone number in case I had any questions as I transitioned to college. These happenstance small acts of kindness are no anomaly among Purdue engineers. This is a culture built by this community that spans across the globe and every decade. And as we lift each other up, we grow to become leaders. All of these individuals demonstrate the Purdue connections spanning the nation and the globe.

Another influence in terms of how Purdue engineering develops incredible engineers is the professors in the College of Engineering. The professors devote time and energy to each student and personal stories, such as Dr. Karen Marais's commitment to my learning experience that have been so impactful. She is a dedicated professor in the School of Aeronautics and Astronautics who first and foremost shows up for her students to support them in all they do. She begins class with a smile on her face and is intentional in learning about her students to get to know us as individuals by asking about the highs and lows of our week. Further, Dr. Marais demonstrates care in supporting her students in their career aspirations. She ensures her availability to schedule time to discuss challenges within the coursework, brainstorm group projects, and support our career development outside of the classroom.

Personally, I have been impacted by Dr. Marais's teaching and have been able to share my career aspirations with her and receive advice on how to best succeed in my goals to continue the learning I have begun in her classes. She has provided strong support for me during my undergraduate career, which is integral for success within Purdue engineering. She makes my education at Purdue more impactful by encouraging me to work hard for my dreams and believing in my career goals. Dr. Marais's support and commitment are invaluable in a professor, because when students feel supported, it is easier for us to persist through the challenges of Purdue engineering and remember the motivation behind our education. Her commitment to her students and all other Purdue engineering professors is what has made Purdue an impactful university to learn at, and this will continue to foster strong future leaders.

Finally, Purdue engineering has demonstrated to me that there is no dream too big to pursue. The supportive community around me has made my achievements possible, because without the support of my family, my friends, my advisers, faculty, and classmates, my aspirations would be intimidating. It is due to their support that I know my strengths and have the courage to accomplish the challenges throughout life while continuing to find new abilities within myself. Every individual I interact with as a friend, a leader, a sister, or a student I hope to encourage as the Purdue communities around me have done throughout my college journey. From happenstance meetings with Purdue alumni and the director of Dayton Ohio Air Force Base, who encouraged me to find a career in aircraft design on next-generation technology and believed in my dreams after one LinkedIn message and phone call, the power of Purdue engineers is strong.

Altogether, Purdue engineering empowers students to share their talents and light with the world, which will without a doubt continue into the next 50 years. It is not enough to be an engineer behind a desk; we must inspire humanity to truly change the world. The persistence, passion, and community fostered at Purdue can move mountains and unite individuals. Time and time again this has been imparted to me throughout my undergraduate career, so as I look ahead to the future, I can state with certainty that no matter what decade it is, Purdue engineers will work together to continue inspiring humanity to work hard on seemingly impossible tasks.

29

PURDUE TREASURES

Reminders of a Legacy of Excellence

BECKY MUELLER

Becky Mueller earned her master's of science in mechanical engineering in 2009 and is a senior research engineer at the Insurance Institute for Highway Safety. She is a Society of Automotive Engineers fellow, a member of the inaugural Purdue Engineering 38 by 38 class, and a recipient of Purdue's Outstanding Mechanical Engineer Award.

Over the last 150 years, Purdue University has acquired many unique treasures that expand the rich culture of the Boilermakers, whether it's the iconic engineering fountain, a mandatory stop during the fountain run tradition, the acoustical engineering wonder of the clapping circle, or the inspiration of the bell tower, representing the pursuit to reach "one brick higher," just to name a few. One lesser-known Purdue treasure holds a special place in my heart: the Apollo 17 moon rock. Located in the Neil Armstrong Hall of Engineering as part of the Roger Chaffee astronaut and Purdue alumnus exhibit, the moon rock serves as a small but poignant symbol of how seemingly impossible goals can be achieved by Boilermakers.

As a student at Purdue, I often visited the moon rock as I waited for friends to finish class in the nearby lecture hall. To me, the rock symbolizes the giant leaps that have been made by other inspirational Purdue graduates. Seeing the moon rock up close reminded me that those seemingly distant goals are closer than I might think, giving me a little extra motivation on tough days to keep pursuing my dreams.

Aiming for the moon, setting ambitious goals, and dreaming of accomplishing things no one had yet dared to attempt has always been a personality trait of mine. As a child, I envisioned myself in a career as a doctor, where I would have a hands-on approach to saving people's lives. Simultaneously, I was passionate about cars. When I was 10 years old I saw the first Insurance Institute for Highway Safety car crash tests on TV, and I knew it was my calling to develop the next generation of crash tests. As a crash test engineer, I could have an exponentially greater impact on people's lives by designing safer cars that protect thousands of people from injury. Honestly, it also sounded like a lot of fun to be paid to wreck cars for a living, and in fact it is!

Purdue gave me invaluable tools and connections to turn my ambitious goals into real-world impact. Specialized courses and research in injury biomechanics enabled me to excel in independent research and contribute significantly to the field. Leveraging Purdue University's strong industry ties, I secured a pivotal internship at General Motors, marking my entry into automotive safety. A chance meeting with a GM engineering manager at a Herrick Laboratory networking event led to this opportunity. The subsequent vehicle safety internship equipped me with vital knowledge in injury biomechanics. This propelled my career as an automotive safety engineer, and within a year of graduation I landed my dream role at the Insurance Institute for Highway Safety. Nearly 30 years after declaring my "moonshot" goal, I humbly find myself being recognized by my alma mater for making giant leaps in my field, as the crash tests I have championed are estimated to save 10,000 lives annually. In 2022 I received the Outstanding Mechanical Engineer Award, and in 2024 I was inducted into Purdue Engineering's inaugural 38 by 38 class among dozens of other alumni making big impacts in the world. While I am so touched by these recognitions, I remain steadfast that my mission will not be complete until there are no fatalities on roadways.

The future of engineering will require speed and agility to pivot to new technologies at a lightning-fast pace. The integration of technology into aspects of everyday life adds complexity and costs but reveals opportunities that were never previously considered. Engineers must be nimble and think outside the box to uncover ways to maximize and integrate artificial intelligence to our advantage into research and development, manufacturing, and the products themselves. To effectively accomplish these goals, engineering teams must embrace diversity of their members, including more interdisciplinary experts and world perspectives to identify problems and propose solutions from a multitude of perspectives. These teams will be led by pioneering seasoned engineers willing to abandon obsolete tradition and by the next generation of engineers educated with future-focused instruction.

Innovative problem-solving and discovery-based education is the foundation for preparing engineers to tackle the unknown challenges of the future. Active teaching methods that involve more practical applications and collaborative interdisciplinary groups allow students to connect book learning to real-world applications. Inspirational teachers with exceptional engineering prowess and experience with cutting-edge research will teach students critical thinking strategies for solving the world's toughest challenges. Day-one access to state-of-the-art laboratories will give students opportunities to solve open-ended research questions, applying engineering principles and formulas to the world around them. Partnerships with industry both on campus in industry-sponsored laboratories and through internships and co-ops will expose students to actual tools and software used by companies, making a seamless transition from student to professional engineer. Research focusing on the most time-relevant consequential global issues such as cybersecurity, cancer research, and artificial intelligence will keep students engaged with their projects. Expansion of study abroad connections with universities across the world enables students to gain different cultural and academic perspectives that will enhance their abilities to tackle the complexities of global companies and customers.

The next generation of the Boilermaker Special is already in motion when it comes to Purdue reconfiguring to the innovative engineering education model of the future. Since I was a student on campus 15 years ago, Purdue has already made dramatic shifts in curriculum, campus life, infrastructure, industry partnerships, and extracurriculars. The Wilmeth Active

Learning Center is a great example of Purdue's future-focused vision of creating a hub for innovative classroom instruction, collaboration, and archives of knowledge. The Gateway Complex of Dudley and Lambertus Halls houses state-of-the-art laboratories including smart factories with robots that use the same equipment and software as industry, giving students valuable research experience and making it easier for them to excel in their careers upon graduation. The Mechanical Engineering building renovations epitomize the vision for the future of learning at Purdue. The renovated space is a complete transformation of traditional offices, lecture halls, and computer labs, bolstering hands-on learning with machine shops and makers spaces that expose students to a broad range of materials and tools for manufacturing plastic, textiles, and metals. This is paired with an overhauled curriculum that prioritizes group- and project-based learning to maximize student engagement. Active learning is everywhere you look on campus: walk through Dudley Hall and look up to see exposed building structures that demonstrate construction principles in action. The Wilmeth Active Learning Center's old powerplant components on display highlight the engineering required for these historical machines, another great reminder that engineering is all around us. The study abroad program covers more than 30 countries, expanding student perspectives to solve future global challenges. Campus life has been revamped in ways that promote a healthy life balance for students with access to a variety of healthy food options, new student housing, and the construction of safe walking and bike paths connecting on- and off-campus students. The 3rd Street pedestrian corridor gives students a break from studies with leisure activities such as cornhole, table tennis, and a piano. Purdue's comprehensive plan for transforming the engineering education process will set future graduates apart, preparing them for the leadership and technical roles they will need to fill to solve the unprecedented challenges of the future.

As for automotive safety engineering, the next 50 years will be marked with the integration of novel and expansive technology focused on vehicle occupants and the surrounding environment. Engineers will use technologies that can identify vehicle-occupant characteristics, including age, weight, health, and alertness status, to provide personalized restraint system protection for occupants when collisions are unavoidable.

To accomplish this, the automotive safety industry must lean on biomedical engineers and medical doctors to expand knowledge of human injury tolerances encompassing the entire span of human characteristics: fragility, obesity, sex, and other comorbidities. Video and other sensor data collected by the vehicle must then interpret characteristics of the occupant and make decisions based on injury tolerances and crash configurations to deploy personalized airbag and seatbelt protections. Integrating artificial intelligence into these decisions will expedite the ability to implement such personalized protection. However, privacy concerns about cameras and sensors capturing biometric data and customer acceptance have the potential to delay implementation. Engineers will need to find nontraditional strategies to overcome these obstacles.

Future vehicles will assist drivers with tasks and prevent many crashes from occurring. Engineers will develop novel vehicle technologies to overcome the biggest driver-related factors of speeding, impaired driving, and lack of seatbelt use that contribute to roadway fatalities. Intelligent speed assistance that pairs with GPS can limit a vehicle's ability to speed, alcohol interlocks can measure blood alcohol levels to prevent impaired drivers from operating a vehicle, and seatbelt sensors can remind drivers and passengers to buckle up, providing increasingly annoying reminders until compliance. Extensions of these concepts can intervene for drowsy and distracted drivers and those experiencing medical incidences and even pull the vehicle off the road, stop safely, and call first responders. Integration of thermal cameras can supplement existing sensors to assist drivers by identifying and stopping for nighttime pedestrians, which could address nearly half of fatal pedestrian crashes. Unconventional pairing of sensors used to monitor the environment with airbag controls could identify impending crashes with pedestrians and bicyclists and deploy airbags to cushion the impact. Future engineers will strive to optimize driver-assistance features to minimize crashes, but crashes cannot be eliminated until drivers are completely removed from the driving.

Autonomous vehicles on roadways will dramatically shift automotive safety research priorities. Roadway transition to full automation will have a "messy middle" phase whereby a mix of fully, partially, and nonautomated vehicles share the roadway. A combination of human drivers and automated vehicles interacting and making decisions is the most complex for crash predictions. Certain common crash scenarios may disappear entirely, while other new unexpected or never before seen scenarios may arise. Over the next 50 years, the engineering required for full automation will be within reach because of the advancement of sensor technology, innovation in cybersecurity, and the integration of artificial intelligence. Whether we reach full autonomy on roadways in the next 50 years will be highly dependent on factors outside of engineering including legislation, infrastructure, environmental change, global political and economic trends, and customer perception and acceptance.

As we explore and inhabit other worlds, automotive safety experts will tackle the unprecedented challenges of extraterrestrial vehicle safety. We will need engineering leaders who can think outside the box to solve problems of predicting crash kinematics in a low-gravity environment, designing inflatable airbag-like cushions and seatbelts to accommodate spacesuit-wearing occupants, and engineering load paths for structures to protect the vehicle propulsion systems from explosion during impact. Referencing knowledge already gleaned from history but finding unconventional ways to adapt to fundamental differences will be key to these future innovations.

In 50 years, Purdue University will acquire many more unique treasures that highlight the giant leaps of Purdue alumni who dared to change the world. Maybe it will be a bronze statue of a pipet honoring the researchers who discovered a cure for cancer, a greenhouse dedicated to those who reengineered crops to improve food security, or a piece of rock commemorating a human's first small steps on Mars—by a Boilermaker.

30

ENGINEERING 50 YEARS FROM TODAY

ELLA

Ella is a fifth grader in Mrs. Lehman's class.

Ella
Mrs. Lheman

Engineering 50 Years from today

In 50 years, engineering will be heavily influenced by technological advancements and global challenges, requiring engineers to be adaptable so this shows that engineers are a important part of history.

In the years to come I think that engineering will be a very important part of history.

31

SHAPING THE FUTURE OF ENGINEERING

MAITRI PANDYA

Maitri Pandya, who holds a B.S. Honors in mechanical engineering (2025), cofounded the Purdue Women in Mechanical Engineering student organization and served as vice president of internal affairs for the Purdue University chapter of the Society of Women Engineers. She was named a 2024–2025 Purdue Engineering Fellow.

E ngineering is the backbone of society. It is what drives advancement and innovation, furthering the technology we use every day to accomplish unimaginable tasks. But what does that really mean? Before college, I never truly understood what it meant to be an engineer. The descriptions I was given never did justice to what I now know the word "engineering" entails. They fell short of capturing the exhilaration of solving a problem that has stumped you for weeks or the thrill of collaborating with others to bring complex and ambitious ideas to life. At its core, engineering is about pushing boundaries and creating solutions that transform the way we live.

When I was young, I considered myself an avid problem solver. My parents often found me dismantling various toys trying to figure out how they worked or constructing makeshift gadgets out of construction paper and duct tape. As I got older, I began channeling these same traits into my newfound hobby of baking. While I wasn't a baking prodigy by any means, it was the scientific nature of the process that truly appealed to me. Testing different recipes, ingredients, and cooking times taught me valuable lessons in trial and error as well as the importance of persistence in achieving the desired result. Gaining a taste of these skills through a hobby I was already passionate about further fueled my motivation for problem-solving. This enthusiasm combined with my continually growing interest in STEM (science, technology, engineering, mathematics) topics throughout grade school inspired me to pursue a career in engineering.

When I first set foot on Purdue's campus as an out-of-state student from Arkansas, I was completely unfamiliar with the environment and didn't know a single other person on campus. Within my first few weeks, what stood out to me the most was an overwhelming sense of community. I met countless other engineering students who shared my passions and drive, and I immediately felt like I belonged. I became deeply involved with women in engineering student organizations on campus, creating and serving as copresident for the Purdue Women in Mechanical Engineering student organization as well as serving as vice president of internal affairs for Purdue's chapter of the Society of Women Engineers. Participating in these organizations strengthened not only my belief in my abilities as an engineer but also my abilities as a leader. I have never considered myself a particularly vocal person, so having the opportunity to create an impact and advocate for others was an experience like no other. It was these experiences that instilled the confidence I needed to become the engineer I am today. I made lifelong friendships through my involvement in these organizations, and they further provided me with a community on campus. This sense of community has stayed with me, revealing how deeply engineering relies on collaboration and how vital these connections are to succeeding in the field. Something I hear all the time is that you can't do this degree alone, and that has proven true to me through my experiences. Receiving support and, more importantly, providing support is a crucial aspect of engineering, and it is something I am grateful for that Purdue instills in its future engineers.

Alongside the community I built at Purdue, my passion for STEM topics flourished into a love for the art of engineering. I gained a deeper appreciation for many of the subjects I studied, with each one helping me make more sense of the world around me. My passion for coding grew, leading me to pursue a minor in computer science. I experienced the same exhil-aration of solving problems that I had in my earlier hobbies, but this time I felt that the ideas and challenges I was tackling held real significance. As I now prepare to leave college and embark on my engineering career, I am eager to apply these passions and values that Purdue has taught me to the larger challenges and opportunities in the field. The possibilities for growth and advancement feel limitless, making it fascinating to imagine what the future of engineering will look like.

Technological advancements have grown exponentially in recent years, spanning a wide range of fields. A few fields that I have gained interest in and am particularly eager to contribute to in my career are automation and sustainability. Both fields have immense potential to shape the future and drive positive change.

Automation has been steadily growing over the past few decades but has seen a significant increase in recent years. The COVID-19 pandemic exposed the vulnerabilities of human-reliant systems due to necessary quarantine and social distancing protocols, further highlighting the need for autonomous technology. There are several industries where we are see-ing a transition to autonomous systems such as medical devices, transpor-tation, and retail systems, among many more. One sector that has been most impacted and one I have gained some knowledge about and experi-ence through various internships is the manufacturing space.

Manufacturing has more recently been undergoing significant trans-formations, driven by a renewed focus on continuous improvement and lean principles across many industries. At the heart of this shift is the in-tegration of automated systems, driven by the goals of minimizing hu-man error, enhancing operational efficiency, and prioritizing workplace safety. One example I had the opportunity to contribute to was a pro-cess of improving ergonomic efficiency in a repetitive hand-drilling process that lacked proper support for body alignment. Implementing an autonomous machine in such instances improves the long-term safety of the operator and also increases precision in certain procedures that are dif-ficult to obtain otherwise. I believe we will see more of these implemen-tations as we continue to increase the efficiency of manufacturing in fu-ture years.

With this growth comes the capability to perform a wider range of operations that are well suited for a machine. Oftentimes, while a design might be workable in theory, it can prove impractical or impossible when translated into manufacturing. Certain limitations that come with human-dependent processes can be reduced through these technological advancements and can be used to continue to bridge this gap between design and manufacturability. The increasing prevalence of artificial intelligence has also been supporting this growth, as it has the potential to increase the scope of what we can create.

Another aspect to consider when discussing the future of automation in society is the way humans interact with these technologies. During my time at Purdue, I had the opportunity to perform undergraduate research at JAIN Research Labs and study human trust with autonomous systems. While autonomous technology has existed for some time, it has only recently gained widespread attention. With this increased visibility, autonomous technology now faces challenges related to unfamiliarity and potential skepticism among the general public.

Research has shown that individuals can be grouped by their trust behaviors based on how they interact with these systems. Ultimately, even the most advanced technology will occasionally experience errors or miscalculations, so it is important that an individual's trust and perception of the technology equates strongly to the reliability of the actual machine. As autonomous technology becomes more prevalent in our everyday lives, this idea of calibrating and adapting technology to fit an individual's trust behavior tendencies will become an important milestone in improving human-machine interactions. Fostering trust and understanding between humans and machines will be key to unlocking autonomous technology's full potential and ensuring its seamless integration into society.

In parallel with the growth of automation, sustainability has been a major focus for several years now, with many businesses striving to reduce their emissions and carbon footprints. Research facilities, in both industry and academia, have been dedicated to numerous sustainability initiatives, including reducing energy consumption and advancing carbon capture technologies. These initiatives have also been scaled to a consumer level, with a rise in products such as electric vehicles and biodegradable packaging. The increased prevalence and knowledge of these climate needs will continue to be the driving force for sustainable innovation. As more industries increasingly acknowledge the importance of addressing these environmental needs, it is my hope that we will see a shift toward embedding sustainability into their daily practices and operations. Treating sustainability as a priority will encourage the development of technologies that not only drive efficiency and profitability but also preserve the planet. With these advancements, we can build a future where innovation does not come at the expense of our environment but instead strengthens the systems that sustain it.

Considering the incredible innovations and technologies on the horizon, it is exhilarating to imagine how they will transform our everyday lives. The exponential growth in technology we have seen in just the last 20 years is difficult to fully grasp, let alone what possibilities lie in the next 50 years. Things once thought impossible are now an integral part of our everyday lives.

Engineering is truly capable of revolutionizing the way we live, work, and interact with the world, pushing boundaries and creating new possibilities. With these new possibilities, I am excited to witness the growth of diversity within engineering and STEM careers as we continue to push for greater inclusivity, unlocking perspectives and innovations from individuals of all backgrounds.

As engineers, we are often tasked with solving the world's most complex problems. But no single person can do it alone. The strength of engineering lies not just in our discoveries and inherent passions but also in the communities and support systems we develop simultaneously. Purdue University has been instrumental in providing me with a strong academic foundation and the support I needed as I began my engineering journey. I am eager to continue expanding my knowledge, making change, and growing my support system throughout my career. I encourage everyone, regardless of their background or field, to cultivate connections and create strong support systems. In the world of engineering, where ideas stretch the boundaries of what we thought possible, it truly takes a village. It is through diverse perspectives and collaborative problem-solving that we turn the impossible into the extraordinary, shaping the future of engineering.

32

CIVIL ENGINEERING IN 50 YEARS

Driven by Technology toward a Sustainable Future

KUMARES C. SINHA AND SAMUEL LABI

Kumares C. Sinha, PhD, PE, Hon. MASCE, NAE, is the Edgar B. and Hedwig M. Olson Distinguished Professor of Civil and Construction Engineering at Purdue University. **Samuel Labi,** PhD, F. ASCE, is a professor in Purdue's Lyles School of Civil and Construction Engineering.

INTRODUCTION

Since the dawn of time, civilizations have built physical infrastructures to address human needs—shelter, potable water, sanitation, and transport. These basic needs have been met by those we call civil engineers. Over the millennia, these needs have not changed fundamentally but have been catalyzed continually by socioeconomic drivers, technological innovations, higher expectations of levels of service, and, in recent years, increased emphasis on sustainability-related outcomes. In 50 years, these trends will significantly transform the civil engineering profession to adapt to a future characterized by smartness (intelligent designs, materials, and processes), sustainability (environmental, social, and economic), and technical efficiency (larger, faster, leaner, and so on).

THE SUBDISCIPLINES OF CIVIL ENGINEERING

Over the years, civil engineering has evolved into a set of subdisciplines as specialty areas. Each specialty area focuses on a particular aspect of civil engineering. Structural engineering involves the design and analysis of load-bearing structures, including buildings, bridges, tunnels, and towers, and structural frames of equipment and vehicles, including ships and air and space crafts. Geotechnical engineering deals with subsurface site investigations, design of foundations for proposed structures, and monitoring and risk assessment of site ground controls. Materials engineering covers the assessment and enhancement of the mechanical properties of construction materials, while hydraulic and hydrologic engineering studies rainfalls, floods, and droughts and deals with levees, channels, canals, dams, and networks for transporting wastewater and water. Environmental engineering focuses on ways to render land, air, and water healthy for habitation by flora and fauna. Geomatic engineering studies feature positioning on Earth's surface to establish man-made and natural reference points and boundaries, while architectural engineering involves building location and configuration planning, design, operation, maintenance, and renovation. Transportation engineering addresses planning, design, operation, and maintenance of highways, pipelines, waterways, railways, and airports as well as pedestrian/cycling facilities, and construction engineering includes contract administration, project control, construction planning and scheduling, and cost monitoring and control. These specialty areas work closely together depending on the nature of the project being considered. In 50 years, boundaries between areas will be more blurred and will perhaps vanish, as societal challenges will demand expertise from multiple disciplines for sustainable solutions drawn from not only civil engineering but also other areas of engineering as well as disciplines from non-engineering areas.

THE PERSISTENCE OF SOCIETAL CHALLENGES AND EMERGENCE OF OPPORTUNITIES

One of the major challenges facing the civil engineering profession is extreme weather events caused by climate change. The future shape of the profession will be driven by how we adapt to climate change and mitigate its effects. Adaptive design principles will be necessary to make civil infrastructures more resistant or resilient to climate change, leveraging artificial intelligence (AI) to provide timely warnings of and effective resolution to climate-induced disasters. Novel intelligent materials will be engineered to withstand temperature extremes and variations that characterize climate change. Coastal and low-lying areas will see elevated or flood-resistant buildings, automated water transport systems, and smart drainage. Also, AI-supported simulation platforms will be used to carry out interactive studies of the prospective behaviors of buildings and other infrastructures under various threat scenarios. Besides climate change, civil engineers will have to contend with other forces on the terrain of infrastructure project development, including the lack of political will and financial constraints, advanced ages and inadequate physical condition of civil infrastructure, uncertainties in the economic environment that exacerbate investment risk, and environmental justice.

THE ROLES OF TECHNOLOGY

GENERATIVE DESIGN AND INTELLIGENT CONSTRUCTION

Generative design will allow civil engineers of the future to simply input their desired technical specifications as either specific standards (e.g., material properties) or general desired outcomes related to technical performance, sustainability, and stakeholder costs and then use AI to develop alternative designs. A typical construction site in 2075 will be expected to feature AI-driven autonomous robots that execute site work including component installation, field guidance for 3D assembly, quality assurance and control of materials and workmanship, and preparation of finished surfaces all at 3D printing construction sites. Automation of the construction phase will also involve unmanned air vehicle–enabled site monitoring, rapid testing of construction workmanship quality and materials, and specification compliance checks all in conjunction with satellite monitoring and remote sensing.

AI-ENABLED CONNECTIVITY AND AUTOMATION TO ENHANCE INFRASTRUCTURE DEVELOPMENT

In 50 years, connectivity and automation will increase the capability of civil infrastructures to exchange data to inform decisions (operational, tactical, and strategic) to enhance process effectiveness and cost-efficiency for the benefit of infrastructure stakeholders. These sibling technologies will be increasingly applied at various phases of development of civil infrastructures. For example, at the planning and design phase of a transportation project where there exists a multitude of alternatives and decision criteria, automation will produce reliable plans and designs, thereby reducing delay and errors subsequently in the project development and delivery.

SMART MONITORING AND MANAGEMENT OF INFRASTRUCTURE

The smart cities concept is characterized by sensor-aided collection of massive infrastructure data, AI-enabled autonomy of infrastructure functions, and Internet of Things (IoT)–enabled connectedness among physical and institutional entities. By the year 2075, advancements in IoT, AI, and data analytics will foster realization of the smart cities concept in human settlements. Infrastructure-mounted sensors will collect and analyze massive amounts of real-time data on internal structural conditions, loading and occupancy, resource (energy, water, etc.) use, and external impacts (air quality, noise, vibration, and so on). Transportation infrastructures including roads, airports, harbors, and railways will become "smarter," with embedded sensors and connectivity devices that help vehicles (of any mode) operate safely and efficiently under all conditions. Infrastructures including buildings, bridges, pipelines, canals, and dams will have smart sensor-based and IoT-connected technologies for self-monitoring and self-reporting of their physical condition and operational performance. This will enable timely response to operational hiccups or sudden defects and timely reactive maintenance in a manner that reduces frequency and cost at minimal inconvenience and cost to facility users and other stakeholders.

DATA SCIENCE AND DATA ANALYTICS

Civil engineers supported by vast amounts of data generated from sensors embedded in roads, buildings, pipelines, canals, and other facilities, will leverage AI to analyze infrastructure internal and performance trends and patterns to simulate or validate complex real-world scenarios. This will help them to optimize infrastructure operations, measure maintenance needs, and enhance user safety and convenience. Also, AI and machine learning models will be used to identify the factors that influence infrastructure physical and performance outcomes in real time, characterize infrastructure relationships with natural and anthropogenic systems, and predict infrastructure behavior under extreme weather, earthquakes, hurricanes, anomalous traffic, and terrorist attacks. This will facilitate the development of resilient and resistant infrastructure designs and management policies in the face of these threats.

SUSTAINABLE DEVELOPMENT OF CIVIL INFRASTRUCTURE

The history of civil engineering over the past millennia has taught us that civilizations that adopted sustainability-related design principles tended to produce infrastructures that were more resilient to the ravages of natural elements. In the coming decades, it is expected that engineers worldwide, in developing scientific solutions to human infrastructure needs, will adopt sustainability as a foundation rather than an afterthought.

With respect to environmental aspects of sustainable development, it can be expected that in 50 years civil infrastructures will use minimal, if any, fossil fuels, with a shift to cleaner renewable energies. In addition, new generations of materials will continue to emerge, including superstrength and lightweight alloys, bioconcrete, carbon-negative and carbon-neutral compounds, and self-healing materials. Buildings will feature green facades, green roofs, net-zero energy facilities, and integrated eco infrastructures that minimize waste. Design innovations will include prefabrication of components to reduce replacement cycles and repair-related downtime. Equity and environmental justice representing the social pillar of sustainable development, will be explicitly considered civil systems planning and design. Overall, civil engineers, as they inculcate technology in their design and management functions, will increasingly ensure that their plans, designs, and operations policies are human-centered and prioritize the desired outcomes related to infrastructure resilience, user safety, and sustainability.

THE EVOLUTION OF CIVIL ENGINEERING PEDAGOGY

In the next few decades, civil engineering curricula will see a significant realignment whereby such subjects as automated sensing, monitoring, and measuring using AI, machine learning, and data science will be increasingly incorporated. It will also become more commonplace for civil engineering students to acquire knowledge in political science, finance, and social science to develop design solutions that are socially, economically, and environmentally compatible with their designs. Also, AI-related technology will influence the way courses are delivered. Virtual teaching assistants will facilitate comprehension of complex concepts and virtual labs equipped with 3D simulations, digital twins, metaverses, and holograms will foster active learning by supplanting or supplementing traditional physical learning infrastructure.

CIVIL ENGINEERING RESEARCH: EMERGING DIRECTIONS

All civil engineering specialty areas will see significant research advancements in terms of the novelty of research inputs, mechanisms and processes, outcomes, and application usefulness. Research inputs will include massive datasets associated with infrastructure design and performance and their relationships with natural and man-made environments and will involve data from wide arrays of sensor types embedded in bridges, buildings, drainage systems, and utility infrastructure. Research mechanisms and processes will include effective management (archiving, organizing, and retrieval) of research data, enhanced visualization (including digital twins and metaverses), machine learning, and AI. Civil engineering researchers in all specialty areas will seek to use this data and research mechanisms to not only produce desirable technical outcomes but also achieve requirements for sustainable development.

Fifty years from now, civil engineering practice will be radically shaped by research advancements. For example, research will allow design flexibility due to computer technology innovations and the increased need for resilience against climate change impacts. Research on geoenvironmental aspects will provide enhanced ways to address natural disasters such as earthquakes and landslides, including cost-effective and rapid repair distressed natural grounds. Research on nanomaterials will render construction materials to be aware of and respond to their environments; capable of carbon absorption; heal themselves of cracks, deformations, and other surface and internal defects; and be lightweight yet possess high strength, plasticity, or other desired mechanical properties. Building research will produce climate-responsive and resilient designs focusing on improved indoor environmental quality and human health productivity. Research in hydraulic and hydrologic areas will produce improved methods of prediction and mitigation of natural disasters such as severe storms, floods, and acute droughts and related events such as wildfires. Massive changes can be expected in the transportation sector due to research advancements in propulsion fuel technology, automation and connectivity, and the use of airborne autonomous vehicles and intra- and intercity drones. Construction will also be driven by advancements in robotics for labor and quality control as well as connectivity between site equipment to facilitate intelligent construction.

CONCLUSION

Unlike products in many other engineering disciplines, a product in civil engineering, such as a building, a bridge, or a transit system, not only requires several years or even decades of planning, design, and implementation but also decades or even centuries. Thus, opportunities for rapid changes in their development in response to changing socioeconomic forces have been rather limited in the past. Advancements in building technology, materials science, and information technology, including modular construction and intelligent materials, have ushered in new paradigms in life-cycle planning and development of civil infrastructures.

The need to accommodate expanding populations in many parts of the world will require new areas for human habitation, and these will include reclaimed land structures, seas, underground cities, and even possible colonies on other planets. This quest will usher in new challenges for civil engineers for decades to come.

In 50 years, civil engineering will evolve into a dramatically different field while addressing the basic societal needs, as it has done over the past millennia. Future practice will be characterized by an integration of the inputs (AI-powered data analytics, automation, connectedness, and smart technologies) and outcomes (infrastructure resilience and sustainable outcomes). Future infrastructures will be not only more intelligent and efficient but also more resilient, sustainable, and responsive to the needs of society and environment and ethical concerns. Civil engineers will work in teams that will be increasingly interdisciplinary in nature, using massive data, machine learning, and AI to optimize plans, designs, and operational and maintenance policies and to predict performance and vulnerabilities of their infrastructures. As the profession continues to evolve, civil engineers will find the need to acquire new skills to ensure that civil engineering infrastructures are resilient in order to continue supporting and enhancing human life in ways that are sustainable and future-proof. In this regard, the ongoing pedagogical changes in Purdue University's civil engineering curriculum are already setting the pace and are rapidly becoming a model for other institutions to emulate.

33

ENGINEERING 50 YEARS FROM TODAY

NAYELI

Nayeli is a fifth grader in Mrs. Lehman's class.

Nayeli Mrs.
Lehman

Engineering 50 years from today

I think that aerospace engineering
and automotive engineering will
have flying cars. In fifty or less
years. Because aerospace engineering
has design, d evelopment, and testing
aircraft. So I think if they both
work together they might have
flying cars.

I think that enviormendal engineering
will make the world have less
a,r and water polution. Because
if the world starts to get so bad
with polution people will start to care.

I think that biomedical
engineering will make lots
of divices for sick people.

34

HOW THE SUPPLY CHAIN-TRAINED INDUSTRIAL ENGINEER WILL POWER THE FUTURE

JAMES A. TOMPKINS

James A. Tompkins, PhD, is an international authority on designing and implementing end-to-end supply chains. He built Tompkins International into a global supply chain consultancy before founding Tompkins Ventures in 2020.

When I studied industrial engineering decades ago at Purdue University, the term "supply chain" wasn't around. In fact, businesses didn't start using the term until the 1980s. For all three of my degrees (BS, MS, PhD), my industrial engineering curriculum focused heavily on efficiency, productivity, and optimizing systems within factories, facilities, and warehousing. Industrial engineers (IEs) such as myself were trained to analyze systems with a focus on how people, machines, and information could work together effectively.

These principles guided my early career and shaped the companies I built. Applying these lessons helped me blaze trails in a field that would soon transform every industry: supply chain management.

Unlike many IEs of that day, I didn't start my career on a factory floor or even a warehouse. After basic training, the US Army assigned me to the Facilities Engineering Department at Fort Monmouth, New Jersey. Then I did a stint as an assistant professor at North Carolina State University. I soon began applying what I taught to my own consulting and design projects, helping companies improve the efficiency of their warehouses and distribution facilities.

Eventually, that became my full-time gig. Over the next few decades, I built a global supply chain consultancy. By 2000, we were helping companies leverage global supply chains as a competitive advantage.

These experiences confirmed that supply chain success, grounded in the industrial engineering principles I learned at Purdue, is the most powerful tool for building resilient, efficient, and innovative organizations.

Today, with everyone's eyes on supply chain after the global pandemic, supply chain professionals are moving from warehousing and shipping to the boardroom table, influencing business strategy. Fifty years from now, I believe that IEs trained in supply chain principles will be leading the world's top companies.

But companies that want to succeed today should not wait 50 years. In a world where disruption is the new normal, supply chain–savvy IEs should be calling the shots—soon.

That's one message at the core of my "retirement" company, Tompkins Ventures, a global B2B matchmaking partnership.

INDUSTRIAL ENGINEERING IS THE CORE OF SUPPLY CHAIN SUCCESS

At its heart, industrial engineering connects the details—optimized parts of a system—to the big picture. The power of industrial engineering lies in understanding how each part affects the whole, recognizing that even small changes can have significant system-wide impacts. When I began my education at Purdue, the emphasis was on making systems work more efficiently, from individual machines to entire facilities. Later on, I fully appreciated how I could apply these principles on a much larger scale.

Let's go back to my first job in the US Army: scheduling Fort Monmouth's 400 tradesmen to handle all repair and maintenance for the fort's facilities. That began a theme that has defined my career and that of many IEs, especially in supply chain: taking on more responsibilities to get the job done.

Maximizing effectiveness required coordinating hundreds of repair and maintenance projects, 1,000-plus work orders, preventive and predictive maintenance, maintenance planning, procurement, inventory management, warehousing, financial management, project planning, and scheduling as well as reporting and creating our monthly Facilities Engineering scorecard.

Interestingly, putting all those parts together provided a great foundation for my future, particularly now, as I use those learnings with Tompkins Ventures to reorient global supply chains.

Industrial engineering helped me see that beyond a maintenance order or a simple warehouse was an entire network. For companies, that global network included suppliers, manufacturing sites, warehouses, and distribution centers. These disparate parts operated better as an integrated whole. I applied these lessons to every part of the supply chain, from sourcing and manufacturing to distribution and customer delivery. That's how I was doing supply chain work before the field even had a name.

Today, supply chain engineering extends beyond company boundaries. It is about optimizing entire networks, often on a global scale. The challenges have become more complex, but so has our ability to meet them.

SUPPLY CHAIN EVOLVES FROM FACTORY FLOORS TO BOARDROOM DOORS

Even after recognizing the term "supply chain," most boardrooms viewed it as a cost center. Its only function was to source and deliver at the lowest possible prices. Finance, sales, and marketing functions drove business growth. Few realized that supply chain strategy could yield competitive advantage.

However, as the business landscape changed, so did perceptions.

From the 1990s to 2020, global supply chains increasingly evolved into low-cost, single-source systems, often centered around China. In this strategy, cost trumped resilience. Multinational corporations leveraged lower labor expenses in developing countries. China offered a vast, inexpensive labor pool and heavy investments in a rapidly growing infrastructure. China joined the World Trade Organization in 2001 and quickly became the "factory of the world."

Companies thrived with just-in-time inventory systems. These systems minimized inventory holding costs but relied on uninterrupted supply chain operations. However, single-source supply chains, by definition, are more vulnerable to disruptions.

And even before COVID-19, the rate and pace of disruption was growing. When COVID-19 shut down the "factory of the world," everybody realized the fragility of supply chains and the risks of overrelying on a single source.

This period marked a turning point. The supply chain was on every newscast, and everyone was an "expert." My wife, Shari, had tried to explain my career to her friends for years. Now, they were calling her up and saying "Supply chain. Isn't that what Jim does?!"

My 31st book, *Insightful Leadership: Surfing the Waves to Organizational Success*, labeled this era the mother of the mother of the mother of all disruptions.[1] In the boardroom, this finally sparked discussions on nearshoring, friendshoring, reshoring, and dynamic optionality to rebuild supply chain resilience. Board members started turning to supply chain pros, even elevating them to the new post of chief supply chain officer.

Supposedly, boardrooms recognize how supply chain performance is the key to profitability. And your chief supply chain officer is as influential as your chief technology officer or chief marketing officer.

But there is still a disconnect, because supply chain and finance types wear different suits and speak different languages. When you look around the boardroom and see gray and blue suits, that's finance, marketing, sales. Then you see the person in the orange blazer and purple pants.

That's your chief supply chain officer. And when your chief supply chain officers start talking about "consolidating LTLs," "FTZs," "end-to-end visibility frameworks," "autonomous actionability," and "demand signal repository," eyes start to roll.

This challenge of language and culture is also an opportunity. Supply chain professionals must bridge this gap.

Looking ahead, IEs trained in supply chain should lead this charge, bringing their systems thinking to the highest levels of corporate leadership. Fifty years from now, supply chain won't just be a function within companies; it will be the nerve center that drives growth, innovation, and resilience.

REGLOBALIZATION: REDEFINING SUPPLY CHAINS IN A DISRUPTED WORLD

This nerve center is the key to success in a world shifting away from traditional globalization toward what I call ReGlobalization. In the post-pandemic era, trade wars, shooting wars, tariffs, geopolitical tensions, and supply chain disruptions make relying on single-source supply chains too risky.

In the Americas, corporate ReGlobalization strategy should take the form of a 3-plus approach. The "3" of Mexico, Panama, and the Dominican Republic are key hubs for production, logistics, and distribution. The "plus" refers to the rest of the world. Industrial engineering–trained supply chain pros can help company boardrooms choose countries by prioritizing proximity, reducing risk, and increasing flexibility.

For markets in North, Central, and South America, ReGlobalization offers faster shipping times, lower costs, and greater resilience to global disruptions. ReGlobalization must incorporate optionality—having multiple sources, suppliers, and routes to adapt quickly to changing conditions.

ReGlobalization is not a full retreat from globalization. Instead, it's a strategic reorientation, creating interconnected networks within regions while maintaining select international partnerships. This approach allows companies to leverage the benefits of globalization while reducing vulnerabilities.

THE ROLE OF INDUSTRIAL ENGINEERING IN FUTURE SUPPLY CHAINS

For a company to succeed today, its entire supply chain network must be resilient. And boards must recognize that their end-to-end supply chain "encompasses every link, from raw material to finished product, no matter which corporate roof that component hangs its hat under."[2] That includes companies' many links upstream and many links downstream from your spot on the supply chain. This potentially involves dozens or hundreds of companies that handle everything from raw materials to final delivery to the end customer.

IEs with supply chain training excel at seeing this big picture: analyzing the flow of goods, identifying bottlenecks, devising cost-saving solutions that don't sacrifice quality, and understanding how each decision impacts the rest of the organization and organizations beyond yours. IEs can guide boardroom choices about where to base production and supply for the greatest benefit. They can develop and integrate the systems needed for truly enabled end-to-end supply chains.

This combination of technical knowledge and systems thinking makes IEs ideal candidates for supply chain leadership and, by extension, the corporate world.

THE NEXT 50 YEARS: AUTOMATION, DATA, AND SUSTAINABILITY

If the past few decades have taught us anything, it's that supply chain is a field of constant change. Over the next 50 years, several trends will reshape supply chain engineering, each requiring the expertise of IEs.

First, automation will become even more prevalent. In warehouses and distribution centers, robots will handle everything from picking and packing to inventory management. This will allow supply chain engineers to focus on overseeing these automated systems, ensuring that they operate at peak performance. By understanding the principles of flow and efficiency, IEs will play a crucial role in optimizing these automated processes.

Second, data will become the backbone of supply chain decision-making. Today, we generate more data than ever before. Each transaction, shipment, and interaction creates a wealth of information. Already, supply chain engineers work with data on an unimaginable scale, using artificial intelligence and machine learning to identify patterns, predict disruptions, and optimize routes in real time. IEs' analytical skills will be essential in harnessing this data to make better, faster decisions.

Third, sustainability will be nonnegotiable. Companies are facing increasing pressure to reduce their carbon footprints, and this trend will only grow. Future supply chain engineers must create networks that are efficient, resilient, and sustainable. We must reduce waste, optimize energy use, and minimize environmental impact. The principles of industrial engineering will play a vital role as engineers develop systems that balance profitability with environmental responsibility.

Finally, supply chain will become an even more integral part of business strategy. As companies recognize the importance of supply chain resilience, the role of the chief supply chain officer will evolve. Fifty years from now, it will be common for supply chain executives to lead entire organizations. They will be at the forefront of strategic decision-making, using their expertise to drive growth, innovation, and resilience.

THE SUPPLY CHAIN IE IS TOMORROW'S CEO

Looking to the future, I am confident that the most successful companies will be led by people with a background in industrial engineering and supply chain. These professionals are uniquely suited for managing the complexities of modern supply chains, which are the backbone of any corporation, particularly multinational conglomerates. Their skills will become more valuable as the field continues to evolve. Fifty years from now, the boardrooms of leading companies won't just include supply chain experts; supply chain pros will sit at the head of the table.

The journey ahead will be challenging, but it will be filled with incredible opportunities. IEs have the skills to make the complex simple, to find the best path forward even when the way isn't clear. For future IEs reading this essay, know that you have chosen a field that will shape the future of business. As IEs, we are poised to lead the way, building supply chains—and companies—that can deliver in a world of perpetual disruption.

NOTES

1 Tompkins, J., & Hughes, M. (2022). *Insightful leadership: Surfing the waves to organizational excellence* (BrightRay Publishing).

2 Tompkins & Hughes, *Insightful leadership* (p. 143).

35

DO WE NEED SOFTWARE ENGINEERS IN 2075?

BOON-LOCK YEO

Dr. Boon-Lock Yeo is a vice president of engineering at Google. He received a BS in computer and electrical engineering from Purdue University and a PhD in electrical engineering from Princeton University and is an IEEE Fellow.

Growing up, I was fascinated with the power of software. During high school, I spent countless hours on the Apple IIe computer, learning programming languages such as BASIC and Pascal, playing adventure games that spanned months, and trying to make programs run a lot faster by coding in 6502 assembly language.

Over a career of almost 30 years I have been working mostly on software, ranging from experience in research labs to leading large distributed teams delivering complex software systems. I have spent many years on software engineering—which I define as the practice of delivering large and complex software systems—working with large teams (usually distributed) and making trade-offs among performance, quality, costs, and features along the way, optimized with limited resource constraints and usually under tight deadlines. Software engineering is much more than programming/coding (the practice of writing codes that computers can interpret).

In recent years, artificial intelligence (AI) has started to change how we work and our approach toward software engineering. We can expect the pace of AI development to pick up in the years to come. Many are already predicting the extinction of software engineers as a career in the era of AI.

Looking forward 50 years to the year 2075, facing the AI challenges, how should software engineering evolve as a discipline? How can software engineers build their careers with the coming age of "real and deep" AI? And how should we educate the engineers of the future?

SOFTWARE ENGINEERING IN THE NEXT 50 YEARS

It is my belief and prediction that software engineering as a practice will be just as important and relevant in 50 years as it is now but will be evolving and in major ways, some of which are the following.

UBIQUITY

Software is everywhere—at work, at home, in cars, under the ground, and in space. If we compare software today and what we had 30 years ago, it is clear that software has made a big difference in improving the quality of our lives in many areas, ranging from software in cars (navigation to assisted self-driving) and software that helps automate our homes (controlling lights and irrigation and monitoring energy consumption) to software in devices that are close to us all day long (smartphones, wearables, headphones, etc.). Most of these did not exist 20–30 years ago. Extrapolating forward, we can expect that software will be much more mature and pervasive—for example, cars would be fully autonomous by then, and different software systems in a car will be working seamlessly with the sensors and devices in the car to ensure a smooth and safe driving experience.

We can also expect software to play a critical role in new usages that we are beginning to explore today, such as in nanobots inside the body for health tracking, sensors under the ground for farming and structure monitoring, robots of all kinds, virtual and mixed reality experience becoming real and prevalent, space travel, and the engine for personal AI assistants.

DRIVING TECHNOLOGY ADVANCEMENT AND INNOVATION

As we see advances in many other areas of engineering and science, including AI, we will need to develop new forms of software and related tools as part of those advancements. Here are a few examples:

- AI as a technology will evolve quickly in the years and decades to come. Software has been and will continue to be instrumental to the realization of algorithmic and architectural breakthroughs in AI. Software engineering will also continue to play an important role in ensuring that AI systems can operate at scale and with high reliability.

- With continued innovations in chip technologies such as faster memory, new interconnect, and extrapolation of Moore's law–type exponential scaling for another few decades, we would require continued innovations from software tool chains, much like how we have seen programming languages and compiler technologies evolve over the years.

- We can expect quantum computing to become a reality within a few decades, and that will demand a whole new set of software to be invented for users to interact with quantum computers.

- Advances in biology in areas such as gene editing, synthetic biology, nanobots that go into human bodies, deeper understanding and prevention of chronic diseases, drug discoveries, and so on hold promises to significantly enhance human health and life spans. Software will play a key role in these discoveries and deployments.

- Software will increasingly serve as close partners to scientists and mathematicians in exploring ideas, making new discoveries, proving theorems, and bringing ideas from different fields to make advancement.

- Another example is space exploration and travel and visiting/living on different planets. Different forms of software need to be invented and deployed to make these a reality.

INHERENT INTERDISCIPLINARY

The field of software engineering will be much more interdisciplinary in nature, with the broadening impact of software. Beyond crafting high-quality software, software engineers would be expected to work more closely with engineers from other disciplines and thus would be expected to have some training in other engineering disciplines. In addition, with many more systems designed by AI, human engineers, working together with nonengineers such as legislators, will play a key role in ensuring that AI systems are designed and used ethically, addressing concerns such as bias, privacy, and safety.

BUGS-FREE

Zero bugs is the Holy Grail of software engineering. High reliability and trustworthiness will be critical when software is operating in demanding environments such as monitoring the heart, driving a car, and operating a nuclear power farm and in many other mission-critical applications. We should strive to develop and certify that a software system can be completely bugs-free and work 24-7 with 100% reliability according to specifications. Bugs will no longer be something software engineers need to ever deal with. Engineers can turn their attention to pushing the limits and speed of innovation.

PERSONALIZED

Software will be designed and configured to the needs of individuals, serving as a personal assistant that understands an individual's needs and also possibly as an extension of one's memory and experience. Education will become much more personal and targeted to one's style of learning than the mass approach we have today. Through collaboration with the personal assistant, we can do more with our time and hopefully live richer and more fulfilling lives.

BRIDGING HUMANITY AND AI SYSTEMS

As we increasingly rely on AI to design software, we need to ensure that such software is built and deployed with proper human oversight and adequate supervision/monitoring/alert mechanisms built in to ensure accurate communications with existing and legacy systems, uphold high standards of safety, and make sure that human values and moral standards are maintained inside the AI "brains." Such work needs to be performed by human software engineers, and it is ultimately most important to ensure that we are responsibly and fully utilizing the power of AI.

PREPARING FOR A FUTURE WITH AI

With the rapid advances in AI, we can expect AI to take over the work of writing the majority of computer codes in the not too distant future. Yet as discussed above, software engineers will still be critical in driving innovations and working with AI systems to design software systems and interpret, validate, and refine AI-generated logic and collaborate with other human engineers and AI systems to deploy and maintain such systems.

As we prepare for a future in which software is thriving and critically important yet dynamically changing, the nature of work that a software engineer will do has to evolve as well in the coming years and decades. For an educational institution such as Purdue University, there are a few fundamental components to be integrated into the curriculum for software engineering (or engineering in general) that would survive the test of time many decades into the future:

- *Mastery of AI and integration of AI into software engineering.* Software engineers will increasingly be expected to master the use of AI for software development efforts. With the advancement of AI, there will be less to no code writing needed eventually, and software engineers need to be able to harness the best of AI tools to design and build software systems. We can also expect software to be developed at a faster pace than is the case today, meaning that the time to develop new features and products can be dramatically reduced. Furthermore, software engineering is more than writing code. There is a strong human element to engineering. (Bill Coughran, a venture capitalist at Sequoia Capital, has famously said, "Engineering is easy. People are Hard.") It will be interesting to see how AI can help soften the "hard" part of engineering through better and more optimized processes.

- *Multidisciplinary knowledge and skills.* With the potential impact of software in many new areas of engineering and sciences, it is important that software engineers of tomorrow be trained in disciplines beyond their major, and they should be equipped with skills to acquire knowledge quickly and with depth, by themselves, and on demand, utilizing the resources available including the assistance of AI. This could mean some extra semesters in college for learning and training so there is sufficient time to cover a broader set of subjects and go deeper.

- *Problem-solving skills.* Systems will become increasingly complex over time, given the advances in multiple areas of engineering and sciences and the utilization of deeper knowledge of specific domains. Being able to quickly diagnose what does not go right, collaborate with the right human or AI experts, and find and validate feasible solutions will be some critical skills that a good engineer needs to develop. It will be important for a university to design courses and projects that expose students to real-world engineering problem-solving instead of just classroom-type problems.

- *Curiosity.* Having a sense of curiosity is the seed to inquiry and new ideas. AI as a tool allows us to do many more things and do them at a much faster pace. It is up to human engineers to harness the power of AI to further push the boundaries of knowledge and invent new technologies. Fostering curiosity early on in one's career would put an engineer at a significant advantage over others.

- *Independent learning skills and lifelong learning.* As we enter an era of increased rapid technology advancement, it is critical that engineers be equipped with independent learning skills, as knowledge learned in schools may quickly become obsolete, and engineers need to learn new subjects quickly by themselves. Developing a habit and joy of lifelong learning will have a compounding effect on what one can accomplish over time.

If I were to start all over again, I would still bet my future and my career on software. I can see myself making an even bigger impact on broad areas of science and engineering via software engineering in the years to come.

Let's go build that exciting future together!

ACKNOWLEDGMENT

The author would like to thank Dr. Minerva Yeung (Purdue BSEE, 1992) for many insightful discussions on the future of software engineering in the age of AI.

36

ENGINEERING IN THE NEXT HALF CENTURY

Enabling, Human-Centric, and Trustworthy

YANNIS C. YORTSOS

Yannis C. Yortsos is the Chester Dolley Professor of Chemical Engineering at the University of Southern California and, since 2005, the Zohrab Kaprielian Dean of the Viterbi School of Engineering.

When I was asked to provide an essay about the future of engineering on the occasion of the 150th anniversary of the founding of Purdue University's School of Engineering, I erroneously thought I would have to predict the future 150 years from now! You see, this is the typical request: *predict the next X years on the occasion of the Xth anniversary.* Wisely, the good people of the School of Engineering, fully realizing that innovation no longer moves on a slow, linear scale, severely constrained their request and asked for a more reasonable prediction over the much shorter period of only 50 years. Or was this a reasonable request?!

Interestingly, for those of us with about half a century of involvement in engineering education and research, when we look back 50 years we tend to view technology and engineering trends and their evolutionary changes as smooth and natural, the paths traced from starting point A and the evolution of their various branches almost deterministic.

Consider, for example, the 2004 seminal report of the National Academy of Engineering (NAE), *A Century of Innovation: Twenty Engineering Achievements that Transformed our Lives*, edited by Neil Armstrong, spanning engineering achievements over the 20th century. Viewed from the present, we can fully justify and rationalize the paths followed that have led to our world today. Nothing very mysterious, right? (And I must hasten to add that Neil Armstrong was an engineering alum of both Purdue [BS] and the University of Southern California [MS]!)

Alas, this hindsight 20-20 vision is unavailable when we try to forecast the future: the nonlinear nature of the underlying systems both technological and social, the countless and inevitable random inputs, the always emerging unintended (often uncomfortable) consequences, and the second law of thermodynamics all preclude a striving essayist, such as this one, from predicting a trajectory anywhere close to the future. Indeed, no matter what half-century segment of the 20th century's technological changes in the NAE report one would have been asked to anticipate, the answers would have likely been far from reality—even for a Jules Verne–like savant! Of course, one can always appeal to science fiction writers or to Hollywood to produce often apocalyptic visions of the future of technology. Or perhaps we can consult a large language model for its vision—and I won't be surprised if such consultation has not already been solicited.

So, there lies the challenge I was asked to tackle. How can one model the evolution of engineering and technological innovation? My approach would be the following: suggest extensive collections of activities, paths, or behaviors, as they are likely to be presented in the other essays in this compendium, and obtain an ensemble average, one that might have some relevance for the future. And if increasing the life span would be one of such upcoming future changes, one hopes that at least some of the present essayists will be around to pass judgment on their colleagues' clairvoyance!

In trying to make sense of such changes, I will go back to my inner chemical engineering, which occasionally tries to poke its head underneath my dean's administrative vestments. With plenty of apologies to economists and innovation gurus, I will make the simple, arbitrary assumption that technological innovation results from a chemical reaction. Moreover, I will denote it (magnitude, extent, potential) by A. Let's first assume that its evolution can be modeled as a first-order chemical reaction, namely

$$A \to A. \tag{1}$$

This autocatalytic mechanism is likely in many innovation processes. Then, with the response being first-order, one obtains the simple kinetics (with a kinetic constant of k)

$$\frac{DA}{DT} = kA \tag{2}$$

that, upon integration, leads to the familiar exponential

$$A \sim ekt. \tag{3}$$

And voilà, Moore's law, but now applicable to innovation at large. Of course, this is very simplistic (and likely problematic). But it reflects that on the aggregate, autocatalysis is fundamental to innovation. Assume, next, a slightly different law, namely that the chemical reaction is second-order. Then,

$$A + A \to 2A, \tag{4}$$

which gives a different kinetic law,

$$\frac{DA}{DT} = kA^2 \tag{5}$$

and the qualitatively much different expression,

$$A \sim 1k(tc - t). \tag{6}$$

Interestingly, a slight change in the kinetics leads to a very different result, a much faster change in innovation that, in the form of equation (6), approaches a singularity (infinity) at time t_c (in theory determined from the initial conditions). And voilà, Kurzweil's singularity theory? Possibly but with a considerable grain of salt! Does the incredible pace of change in artificial intelligence (AI) today, resembling faster than exponential growth, bring about such a regime? Perhaps, but only in a qualitative sense. Indeed, these simple (simplistic?) arguments show that a singular behavior is possible (and I should note that any reaction of order greater than one will also lead to a similar behavior).

When even politicians talk about *exponential* changes, one cannot but accept the thorough, permeating impact of technology and engineering on our world. And I can't wait for the vernacular of *singularity* to enter the everyday political conversation! Is it possible, then, to postulate that today and in the foreseeable future, the technology and engineering world will consist of segments of exponential growth of different rate constants, interspersed with segments where growth is of the singularity type, and that *inflection points* (another favorite utterance of politicians) will regulate the transition from one such segment to another? With appropriate curve fitting, surely much is possible. But even though these are likely to prevail, how can they help us predict the future of engineering in the next half century? However elegant and likely, they do not provide us with a priori predictive power, although they will help us with forensic past analysis. Then, the question reverts to what, if any, remains constant as engineering and technology advance? For this, we can go back to the NAE.

Following its 2004 report, the NAE undertook another study, now to address the *forward* problem, so to speak, one not very dissimilar to the subject matter in this essay, namely what new technologies and engineering products will likely emerge in the 21st century. Eventually, the NAE committee migrated to answering a different question: What are the critical grand challenges for engineering in the 21st century? Chaired by its president Chuck Vest, the NAE issued a 2008 report outlining 14 such grand challenges. They varied across many domains but were all characterized by *complexity* and *scale*. Today, 17 years later, significant progress has been made toward resolving some of them (e.g., *making solar energy economical* and *advancing health informatics*). Some other of the challenges identified (e.g., *providing energy from fusion*) are still unmet. But to this essayist, the most important outcome of the 2008 report has been the inference that all challenges in engineering fall into four buckets of endeavor:

sustainability, health, security, and *enriching life,* roughly mapping to the Maslow hierarchy of needs now applied to society at large.

In a fascinating and empowering confidence in the ability of engineers to solve problems and challenges at scale and complexity, the NAE thus outlined key important areas for our attention and summoned the engineering ingenuity to come up with their solution, in paths that will surely involve exponentials, singularities, and inflection points! Very recently, and a somewhat different incarnation of the NAE grand challenges, the National Science Foundation launched the Engineering Research Visioning Alliance, commissioned to address how engineering can address some related challenges. So, this makes my view of where engineering is headed in the next half century straightforward. We will witness the constant discovery of new materials, devices, energy sources, and technologies at large that will aim to solve the various multiple and constantly emerging questions in the four buckets of *sustainability, health, security,* and *enriching life.* How we will organize ourselves in terms of education, industry, and government, within different nations or within the global community, to address these issues will remain to be seen.

I must add a few relevant points:

1. These four buckets are interconnected and interpenetrating (e.g., sustainability affects health and security, security and health affect enriching life, etc.). They are encroaching on other nonengineering disciplines, including the sciences, medicine, policy, and human-centric disciplines. Thus, we are likely to see the emergence of new multidisciplinary and transdisciplinary areas of specialization where different current disciplines find an increasing overlap (or Venn diagrams, to use one favorite term of public rhetoric today) as in, for example, engineering and medicine. Of note, transdisciplinary fields have been multiplying over the centuries and increasingly so in the last 50 years: chemical engineering (fusion of chemistry and engineering), computer engineering (a fusion of computer science and electrical engineering), etc. One would surely see new such areas developing as convergence dramatically accelerates.

2. Advanced computing (AI, machine learning, and soon-to-be-applied quantum computing) and embodied AI (e.g., robotics and automation) would be powerful means to help cut the various Gordian knots,

particularly as they relate to scale and complexity. *AI+X* will be the new incarnation of Engineering+, a concept articulated several years ago but one with increasing power and impact. Can we claim that not too long into the future all disciplines will be empowered by engineering (in the newer manifestation of AI and related computing technologies), just like all companies are today technology companies?

3. Technological and engineering innovation flourished, at least in the last century, by empowering individual ingenuity, experimentation of new ideas, open competition and markets, and open research. Technological innovation led to evidence-based policy. Now, some of the challenges in the four buckets (e.g., sustainability) would require the reverse, namely innovation driven by evidence-based policy. This also requires global cooperation. By contrast, others (e.g., national security) would increasingly require innovation in exactly the opposite direction, one not globally shared. How we as humanity navigate these challenges will surely create another defining feature of engineering (and possibly a new transdisciplinary field) in the next 50 years.

4. In my final point, I will borrow from an editorial I recently wrote in the USC Viterbi School of Engineering's *Dean's Report 2024*:

> Theodore von Karman famously defined engineering by referring to its extraordinary impact: *scientists discover the world that exists, engineers create the world that never was.* With an ever-accelerating rhythm many decades since his passing, history has demonstrated the profundity of his statement. Today, and in the years to come, we witness the potential of its even stronger manifestation. Singularly talented historian Noah Yuval Harari has captured this power in his intellectually audacious book *Homo Deus*, one of his enlightening series of thought leadership treatises. Von Karman's view of a *world that never was* is about to unfold.[1]

Engineering's exceptional power to shape humanity's even nearer future brings up a gigantic moral demand to all of us who are involved in the formation of the next generations of engineering graduates. This almost existential demand requires us to help endow them with *trustworthiness*:

the grand sum of *outstanding technical competence and outstanding character*. In the past, engineering education concentrated primarily on technical competence. Today, the tremendous autocatalytic rate (exponential or even faster) of the increase of the power of technology requires that our students should also be trusted for their outstanding character.

As I close this brief essay, I would note that the Greek etymology of the word "character" comes from "etching," suggesting fundamental integrity. It is fascinating that a crucial current topic for AI research is *trustworthy AI*, which in many ways alludes to the etching ingrained in its software. But I would venture to endow this element of *trustworthiness* to all technology and engineering, not simply AI, and combine it with the purpose that our engineering graduates should have to help our vision to *engineer a better world for all humanity*, a task more urgently needed than ever before. And this would likely be another defining element of engineering in the next half century.

NOTE

1 USC Viterbi School of Engineering, *Dean's report 2024: Trustworthy engineers; Restoring trust in untrusting times*, https://viterbischool.usc.edu /wp-content/uploads/2024/12/USC-Viterbi_Deans-Report-2024.pdf.

ABOUT THE EDITOR

Arvind Raman is the John A. Edwardson Dean of the College of Engineering and the Robert V. Adams Professor of Mechanical Engineering at Purdue University. He is internationally recognized for his research on vibrations and nonlinear dynamics in nanotechnology, biomechanics, and manufacturing. Raman has been instrumental in founding the Shah Family Global Innovation Lab in the College of Engineering, which fosters technology development and translation of technologies for sustainable development. His notable contributions include advancements with the atomic force microscope, an indispensable tool for scientists and the industrial community to better image and measure properties of complex materials. Additionally, simulation tools developed by Raman's group on the nanohub have been used by thousands of researchers worldwide. An ASME fellow and recipient of the ASME Gustus Larson Memorial Award, a Keeley Fellowship, a College of Engineering outstanding young investigator prize, and an NSF CAREER award, Raman has demonstrated a lifelong dedication to engineering innovation.

www.ingramcontent.com/pod-product-compliance
Lightning Source LLC
Chambersburg PA
CBHW070916270326
41927CB00011B/2602